Algorithms, Monks, and Merchants

Computing in Everyday Life in the Middle Ages

Algorithms, Monks, and Merchants

Computing in Everyday Life in the Middle Ages

Giorgio Ausiello

Sapienza University of Rome, Italy

World Scientific

NEW JERSEY • LONDON • SINGAPORE • BEIJING • SHANGHAI • TAIPEI • CHENNAI

Published by

World Scientific Publishing Co. Pte. Ltd.

5 Toh Tuck Link, Singapore 596224

USA office: 27 Warren Street, Suite 401-402, Hackensack, NJ 07601

UK office: 57 Shelton Street, Covent Garden, London WC2H 9HE

Library of Congress Control Number: 2025001876

British Library Cataloguing-in-Publication Data
A catalogue record for this book is available from the British Library.

Original title: *Algoritmi, monaci e mercanti. Il calcolo nella vita quotidiana del medioevo,
by Giorgio Ausiello*
Copyright: © 2022 Codice edizioni, Torino
Cover image: Images of al-Khwarizmi, Luca Pacioli, Leonardo Pisano, and digital calculus

ALGORITHMS, MONKS, AND MERCHANTS
Computing in Everyday Life in the Middle Ages

ISBN 978-981-12-9927-8 (hardcover)
ISBN 978-981-98-1283-7 (paperback)
ISBN 978-981-12-9928-5 (ebook for institutions)
ISBN 978-981-12-9929-2 (ebook for individuals)

For any available supplementary material, please visit
https://www.worldscientific.com/worldscibooks/10.1142/14019#t=suppl

Typeset by Stallion Press
Email: enquiries@stallionpress.com

Preface

Algorithms have become an extremely popular subject in recent times, and their role in our daily lives is a topic of wide-ranging discussion. The widespread presence of the web and computer systems, which constantly follow us and support us in our various activities, and the invisible presence of algorithms, which simplify our search for information, advise us on our online purchasing choices, and guide us in our financial investment decisions — to give just a few examples — have made us dependent on algorithms in our daily lives, even though we may not always be aware of them.[1] As a result, we have been overexposed to the term, and concept of, algorithms in the press and media, where they are often applied inappropriately. The use of algorithms is sometimes overemphasized as an effective approach to information management and problem-solving; at other times, however, it is demonized and offered as an example of the way we delegate important decisions to abstract, impersonal, and ultimately unaccountable computation and decision-making mechanisms. A mistrust of the use of algorithms in decision-making processes is especially felt in the law, in fiscal matters, and in educational assessment.

In actual fact, however, if we analyze documents on the use of mathematical computation and algorithms over the course of human history, starting at least two or three thousand years before our times, we will see that the need to compute and solve mathematical problems by using

---

[1] On the presence of algorithms in contemporary life, see Giorgio Ausiello and Rossella Petreschi (eds.), *The Power of Algorithms*, Springer, Berlin, 2013.

appropriate computational rules has always accompanied human life; indeed, it was specifically everyday human needs that led to the birth and development of mathematics and computation.

Based on these considerations, the idea arose to present in this book the close relationship that existed between application-oriented requirements and mathematical problem-solving in an important period in human history — the Middle Ages, particularly between the 9th and 15th centuries. This period was a time of great cultural evolution, beginning when Arabic science was in its most flourishing phase, continuing with the acquisition of Indo-Arabic calculation methods based on the decimal positional system in Western Europe, and ending with the rediscovery of Greek cultural and scientific sources, which were first handed down in Arabic and later in Latin translations.

I should make it clear at the outset that the objective of this book is not to present a detailed history of algorithms, much less a history of mathematics in the Middle Ages. My intention is to illustrate a selection of significant examples to show how mathematical computation was inextricably linked to the solution of the problems people encountered in their daily lives in the Middle Ages: from the religious question of calculating the date of Easter to the measurement of land, calculations of volumes of various kinds (wells, cisterns, or city walls), distribution of grain and other agricultural products, calculations of interest, and solutions for problems relating to currency exchange. Especially in the late Middle Ages, which experienced a remarkable economic revival, it was the activities of merchants that made computational skills an essential requirement for solving a wide variety of problems; it was this that led to the creation of the so-called abacus schools that the children of merchants — first and foremost from Tuscany but also from various other Italian and European regions — attended in their teens in order to learn the calculation techniques they needed to carry out their business activities profitably.

Before going on to illustrate the spread of algorithms in the Middle Ages, I believe it is appropriate to briefly mention their presence in the ancient world in Chapter 1. The expertise in mathematics and algorithms of the second and first millennia BCE that was acquired in Mesopotamia and Egypt and the computation techniques developed in China and India have retained their value through the centuries and left their mark on the history of mathematics. One example of this is the sexagesimal notation we still use today when we measure time and angles. As we will see in Chapter 1, it is precisely the everyday actions of humans (land management, astronomical observations, buying and selling products, and dividing inheritances, for example) that have guided the definition of problems and solutions since antiquity and led to the birth of entire fields of mathematics such as geometry, algebra, and trigonometry.

Chapter 2 is devoted to Arabic mathematics and, in particular, the work of al-Khwarizmi, who plays a central role in the development of this book because of the major influence Arabic computational mathematics has had on the European West — to the point where al-Khwarizmi's name (albeit mangled to become 'algorism' or 'algorithm') has been attributed to every detailed computational procedure.

Chapters 3 and 4 seek to show how — in a Europe that was still tied to a numerological tradition with religious influences, and at a time when the concept of computation was essentially applied to the calculation of the date of Easter and calculation skills were mostly cultivated at court and in monasteries — translations of Arabic manuscripts by Jewish mathematicians and Iberian monks revealed the power implicit in Indo-Arabic systems of calculation to the Christian West. These methods of calculation were somewhat slow to penetrate, however, and we have to wait until the 13th century to see them being established in everyday life in the European world, with the publication of *Liber abaci* by Leonardo Pisano, who is better known today as Fibonacci.

Leonardo's work, which I discuss in Chapter 5, is emblematic from several perspectives. First of all, the subjects it covers mean that it is the fundamental link between Arabic mathematics and the Western world. Second, because of Leonardo's particular attributes (he was the son of a Pisan merchant and a merchant himself) and the context of the city he lived in, his book provides an indication of the relationship that would tie mathematical calculation and the mercantile world closely together for over three centuries. *Liber abaci* is therefore a pillar of the history of calculus, and all the didactic contributions of hundreds of 'abacus masters' over the centuries that followed drew inspiration from it. These texts, which we discuss in Chapter 6, were tackled by thousands of young men, mostly the sons of merchants eager to be trained in the techniques of calculus, which was a fundamental tool if they wanted to establish themselves in the Italian and European markets.

Chapter 7 is devoted to the evolution of abachistic mathematics in the 15th century and the emergence of the relationship between mercantile mathematics and the humanistic world. This chapter brings our review of the role of algorithms in everyday life in the Middle Ages to a conclusion. It is precisely in the last part of the 15th century, and especially in the 16th, that a number of factors — the rediscovery of classical Greek and Latin texts by scholars, the invention of printing, and the resumption of a more abstract approach toward mathematics that was less closely tied to the material needs of the mercantile rationale — led to changes in the scenario of European mathematics, and the role of computational methods and algorithms became less important until their forceful reemergence with the advent of electronic computers.

Some important aspects need to be considered before I conclude this preface. The first relates to the concept of 'algorithm'. Our modern culture has accustomed us to defining an algorithm as a sequence of clearly specified computational steps that can be carried out by a human being or a computer with no ambiguities and (if possible) with guaranteed termination. The algorithms we will encounter in this book only exhibit these characteristics in some cases. For example, some of the learning exercises for student scribes we find on Babylonian tablets reflect a repetitive and highly detailed style that can be compared to that of a computer algorithm. In other parts of this book, however, when we look at the abacus books produced between the 13th and 15th centuries, among other things, we will see that problem-solving was presented in a rhetorical, conversational, and non-schematic style. On closer inspection, however, we see that in these cases, too, the description of the steps to be taken was no less precise and no less detailed. Repetition of the solution to the same problem but using different data makes it easy to extract a process from the text that we can unquestionably define as algorithmic.

The second observation concerns the educational writing style employed in this book, which it is hoped will make it suitable not only for readers who are interested in the development of mathematics and calculus but also for those who more simply wish to gain an understanding of the cultural atmosphere of a period of history that witnessed a wild social and economic evolution, first in the Muslim world and then in Western Europe. To make it understandable for readers who are unfamiliar with mathematical texts, I have chosen to take relatively simple examples and problems from the works I cite. In any case, a normal secondary school level of mathematics should allow everyone to follow the topics covered in this book without difficulty and also, I hope, appreciate them.

I should make one final observation to draw the reader's attention to a particular aspect that is closely related to the content of this book and is also of considerable historical relevance. The material that was produced over the course of three centuries — the hundreds of treatises written between the 13th and 15th centuries to prepare young people to use the computational techniques required in everyday life, the so-called 'abacus books' and 'algorithm books', to which several chapters of this book are devoted — represents a heritage of enormous cultural interest. It shows that the symbiotic relationship between algorithms and the needs of daily life must be explored to understand the evolution of computational techniques; however, there is also another possible interpretation. It is from these same books on algorithms and abacuses that we can obtain a detailed — and perhaps even minutely detailed — portrayal of merchants' business activities, the types of food that was eaten, the goods that were exchanged in markets (such as agricultural products, metals, animals, and spices), the fairs where trade was conducted, the currencies that were used for business transactions, the most common kinds of craft, and so on. Although it is not possible to develop this aspect in depth in this book, I believe that the description of the documents and the examples I have provided will be sufficient to stimulate a desire in my readers to understand the links that established the relationship between the material civilization and mathematics in medieval times.

Acknowledgements

I would like to thank my wife Anna Unali, whose research and writings on travel and trade in the Middle Ages have transmitted to me a curiosity about and interest in this fascinating period of human history. I also warmly thank my colleagues and friends Giorgio Gambosi and Luigi Laura for their valuable suggestions.

Contents

Chapter 1

Algorithms and Daily Life in Antiquity

Mesopotamia

The birth of writing and the emergence of mathematical computation were intrinsically linked to the requirements of everyday human life. The earliest forms of written records on tablets, small objects, and clay containers are associated with more ancient periods of agricultural and pastoral activity, and it was with the emergence of cities, and the development of urban organization and bureaucracies that depended on monarchs, autocrats, and priests in particular, that it became necessary to keep accounts of the vast quantities of goods and products flowing into warehouses or being exchanged with neighboring states. It was in this same context that public works of enormous importance — such as the construction of walls, palaces, and warehouses and the digging of canals for irrigation purposes — were carried out. This all required a very large number of officials and scribes to be trained who would be able to perform mathematical calculations and invent innovative ways of solving progressively more complex problems. The oldest evidence of this development includes documents (mostly in the form of clay tablets and papyri) from Mesopotamia and ancient Egypt, which not coincidentally were where some of the earliest civilizations were established and where some of the most ancient urban settlements were founded.

The earliest Mesopotamian tablets were written in cuneiform script and date from the Sumerian and Akkadian periods (2500–2000 BCE).

They offer a detailed description of the mathematical procedures required for problem-solving and attest to the existence of a real training system for scribes. For example, there are the well-known tablets discovered at Shuruppak in the Euphrates Valley south of Baghdad, which date from 2500 BCE and are now housed in the Istanbul Archaeological Museums. At the time, the Sumerians used an additive (that is, non-positional) number system that was decimal and sexagesimal at the same time. The tablets deal with issues relating to the division of resources. In particular, one problem that comes up repeatedly[1] concerns calculations of how many people can be given a share of grain (wheat or barley) if it is known that the granary has a given capacity C and that each person must receive an amount x of grain.

The sexagesimal positional notation system became established in the Paleo-Babylonian period, beginning in around 2000 BCE (Figure 1). The notation is always mixed because the 59 digits needed to represent numbers in a base-60 system are represented in an additive decimal notation. For example, if we separate the various digits with a semicolon, the number 35;4 corresponds to $35 \times 60 + 4 = 2{,}104$. The value 35 is not represented in decimal positional notation but in an additive form $(3 \times 10 + 5 \times 1)$. The zero was not represented, and to indicate that one of the powers of 60 was not present, a space was left, which could naturally give rise to ambiguities that might possibly only be resolved by the particular context.[2] The same sexagesimal notation was used to represent the negative powers of base 60.

The available documentation from this period onward shows the use of mathematics with a wealth of operations and concepts.[3] Among the operations that scribes taught their students, we find addition, subtraction, multiplication, and calculation of the inverse.[4] It should be noted that the

---

[1] Jean-Luc Chabert, *Histoire d'algorithmes. Du caillou à la puce,* Belin, Paris, 1994, pp. 13–14.

[2] Even when the use of a circle to represent zero was introduced in the Seleucid period, it was only to indicate the absence of a power of 60 between two non-zero digits and not the zeros at the end, as in the case of the number 1,300, for example.

[3] Karen Rhea Nemet-Nejat, *Cuneiform Mathematical Texts as a Reflection of Everyday Life in Mesopotamia,* American Oriental Society, New Haven, CT, 1993, pp. 10–15.

[4] Given the number x, the inverse (or reciprocal) of x is the number $y = 1/x$ so that $xy = 1$.

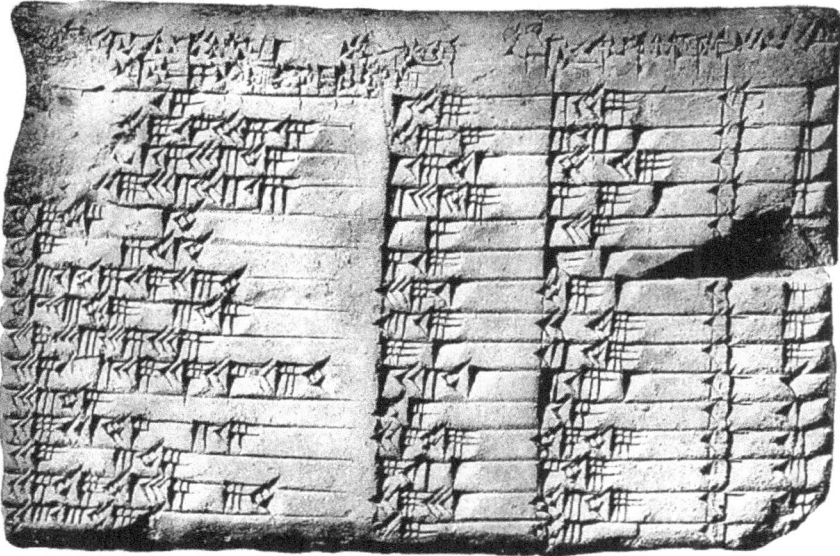

Figure 1. Plimpton Tablet 322 (ca. 1800 BCE). It is named after George Arthur Plimpton's collection and is kept at Columbia University. According to Otto Neugebauer's interpretation (*The Exact Sciences in Antiquity*, 2nd Edition, Munksgaard, Copenhagen, 1951), it contains a sequence of Pythagorean triples in sexagesimal notation.

calculation of the inverse was particularly important because division between a number x and a number y was accomplished by multiplying x by the inverse of y. Operations to calculate the square and the square root of a number (by using approximation formulas) were also known. Finally, it also appears that the procedures for solving first-degree equations and some types of second-degree equations were known.

Plane geometry was applied, sometimes in an approximate fashion, to calculate areas of land in the shape of quadrilaterals, polygons, and circles by using tables of coefficients (sometimes, approximately, for example, the value 3 or in some cases the more accurate value of $3 + 1/8 = 3.125$ was used for π). Plane geometry was also used to make algebraic calculations (by interpreting the product as the area of a rectangle). Curiously, the measurement of angles in the modern sense was not known. This is somewhat strange if we consider that when we measure angles, we use a sexagesimal system, dividing the round angle into 360 degrees. Although it was less well developed, solid geometry was also used to solve practical problems (for example, to calculate the volume of a moat to be dug or the number of bricks needed to build a wall).

Tables were created to avoid performing the same operations repeatedly and were in use from the Sumerian period: There were multiplication tables and tables of inverses,[5] squares, square roots, and geometric series — that is, successive powers of particular numbers — among others. Geometric series (which were used, for example, to calculate how much the capital reinvested at an established interest rate would yield after a certain amount of time) are especially interesting because the inverse concept of logarithms, which involves calculating how many multiplications of a number by itself are needed to reach a given value, is related to them. In a sense, the concept of logarithms was also known to Mesopotamian mathematicians since logarithmic tables (often in base 2) are also found in the available documentation.[6]

Overall, however, the approach to mathematics was essentially practical. One interesting study analyzed a total of 180 tablets of cuneiform mathematical texts from various eras, but predominantly from the Paleo-Babylonian period.[7] More than half contain problems of an algebraic or geometric nature and 75 deal with practical issues. About 40 per cent of the tablets contain only one problem, while a majority, about 60 per cent, include between two and several dozen. The study identified more than twenty types of applications that the problems assigned to the student scribes relate to: brick buildings, digging foundations, canals and irrigation, defensive works, calculations related to trade and purchase and sale agreements, calculations of interest, calculations of inheritances, the construction of hourglasses, raising and managing livestock, metalworking, dividing tasks among workers, and so on.

---

[5] Sometimes, if the context was clear, it was assumed for the purpose of simplifying the calculation that the inverse or reciprocal of x was y if $xy = 60$; y could then be scaled by adopting an appropriate power of the base 60. For example, 4 was considered to be the reciprocal of 15 because $15 \times 4 \times 60^{-1} = 1$.

[6] The term logarithm was coined in the 17th century by the Scottish mathematician John Napier, who studied and analyzed the properties of this function.

[7] Nemet-Nejat, *op cit.*, p. 24.

As an example, let us look at some of the simple problems reported in the study (it should be recalled that we use a semicolon to separate the various digits of a number written in sexagesimal positional notation):

- If $1;1;1;1$ [$= 60^3 + 60^2 + 60 + 1 = 219.661$] sheep are allocated to $13;13$ [$= 13 \times 60 + 13 = 793$] shepherds, how many sheep will each shepherd's herd have?
- Find the amount of bitumen needed to cover a certain area with a one-millimeter-thick layer.
- If a weaver can weave $0;20$ [$= 20 \times 60^{-1} = 1/3$] cubits a day, how long will it take him to weave 48 cubits?
- Calculate the profit made if quantity x of a commodity (oil or lard, for example) is purchased at a price y in silver and if a quantity z of the same commodity is resold at a price w.
- Given a principal amount of money of value x, and given the value of interest y, if the money were lent at a compound interest rate of 20 per cent per year (a year is assumed to be 360 days), how long was the loan duration?

Among the particular problems that give us an interesting glimpse into the customs of Mesopotamian societies (in the historical period corresponding to that of the available tablets) are those regarding the division of inheritances, from which it is possible to observe a patrilineal organization and a strong bias in favor of older male sons: For example, the first and second sons received twice as much as the third and fourth sons, who in turn received twice as much as the fifth and sixth sons. The problem was easy to resolve in the case of bequests of cash or silver, but when it was a question of dividing a rectangular or trapezoidal piece of land, it became necessary to employ out-of-the-ordinary geometric techniques. The problems relating to the calculation of interest also shed light on singular aspects of the economic life of the time; today, a 20 per cent per annum interest rate on loans, for example, seems to us to be far too high for an agricultural society with limited trading activities.

An examination of the tablets reveals that the Sumerians and Babylonians did not prove theorems. Scribes would provide the solution to a problem, but they had no arguments to support whether it was correct, although this could sometimes be verified by requiring the student to perform the inverse calculation. For example, one might be asked to calculate the time t required to accrue a certain amount of interest x at a given rate r and then calculate the accrued interest x given the rate r and the elapsed time t.

An actual concept of algorithms was not a part of the training of scribes, but it is interesting to note that an implicit concept can be inferred from the texts of the tablets. "They [the Babylonians] did not have an algebraic notation that is quite as transparent as ours; they represented each formula by a step-by-step list of rules for its evaluation, i.e., by an algorithm for computing that formula. In effect, they worked with a 'machine language' representation of formulas instead of a symbolic language. [...] Note also the stereotyped ending, 'This is the procedure,' which is commonly found at the end of each section on a table. Thus the Babylonian procedures are genuine algorithms, and we can commend the Babylonians for developing a nice way to explain an algorithm by example as the algorithm itself was being defined."[8]

Let us consider the VAT 6505 tablet[9] (which is kept at the Vorderasiatisches Museum in Berlin[10]), which describes the method for calculating the inverse of a number. As always, the calculation was repeated many times using different numbers, but only five of these examples are readable. For the sake of simplicity, we will only report one of them and will show how the corresponding algorithm can be deduced from it.

The computation is based on the expression $1/n = 1/(y + z) = 1/y \times 1/(z/y + 1)$ and consists in reducing the computation of the inverse of a number n to the computation of the inverses y' and u' of y and $u = z/y + 1$, which are both smaller than n, until numbers are reached whose inverse is available in precomputed tables. The text of the exercise presented in the tablet is as follows (the explanations are in brackets):

[8] Donald E. Knuth, Ancient Babylonian Algorithms, in *Communications of the ACM*, 15, 7 July 1972, pp. 572–673.
[9] Chabert, *op cit.*, pp. 17–19.
[10] VAT stands for Vorderasiatische Abteilung Tontafeln, the Middle Eastern Clay Tablets section of the Vorderasiatisches Museum in Berlin.

The number is 4;10. What is its inverse?

Proceed as follows. (Break down $n = 4;10$ into $y = 10$ and $z = 4 \times 60$)

Calculate the inverse of 10. You obtain 6. (6 is thus the value of y')

Multiply 6 by 4. You obtain 24.

Add 1. You obtain 25. (25 is thus the value of u)

Calculate the inverse of 25. You obtain 2;24.

Multiply 2;24 by 6. You obtain 14;24.

The inverse is 14;24. This is the way to proceed.

In decimal notation, this involves finding the inverse of the number $4 \times 60 + 10 = 250$: that is, 0.004. According to the above-provided formula that was used by the scribe, the result is obtained by multiplying the inverse of 10 (which is 6×60^{-1}) by the inverse of 25 (which is $2 \times 60^{-1} + 24 \times 60^{-2} = 0.04$). It is clear that the following algorithm can be easily derived from the procedure in the numerical example:

Given the number n, let y and z be two numbers so that $n = y + z$.
Calculate the inverse of y. You obtain y'.
Multiply y' by z. You obtain v.
Add 1. You obtain u.
Calculate the inverse of u. You obtain u'.
Multiply u' by y'. You obtain n'.
The inverse of n is n'.

This example shows clearly how the numerical calculations presented to students actually included all the detailed instructions needed to perform the same calculations on other numerical data.

Egypt

Mathematics and computation in ancient Egypt were also motivated by practical reasons. It is a well-known fact, for example, that Herodotus attributed the invention of geometry to the Egyptians in connection with the need to measure land after the Nile flooded. "They said that this king distributed the land among all the Egyptians, giving each an equal lot in the shape of a square, and that according to this division he obtained revenues, having imposed the payment of an annual tribute. If the river took any part away from a farm, the owner went to the king and reported the incident to him. He then sent officials who observed and measured how much smaller the land had become, so that in future, the landowner paid the tribute proportionally. We believe that geometry originated as a result of this and then moved into Greece, and the Greeks knew about sundials, gnomons and the twelve parts of the day from the Babylonians."[11]

As we can see, Herodotus acknowledged the primacy of mathematics as it related to astronomy to the Babylonians, but at the same time, he identified the Egyptians as the "inventors" of geometry because of the need to recalculate the size of lots of land for tax purposes after the Nile flooded. In particular, *harpenodaptai* (literally "rope-stretchers") were responsible for measuring land and making the relevant calculations and therefore received an education in geometry.

Unfortunately, unlike in Mesopotamia, where the use of clay tablets to record documents helped with their preservation, because records in Egypt were kept on perishable vegetable matter, which above all was not fire-resistant, very few mathematical texts have survived that might tell us about the problems that were faced most often and the mathematical techniques used to solve them. What we do know is therefore essentially based on three documents.[12] Two are in the British Museum in London, a leather scroll and a papyrus (known as the Rhind Papyrus from the name of its first owner, two fragments of which are available). The third is a papyrus in the Pushkin State Museum of Fine Arts in Moscow (it is known as the Moscow Papyrus or Golenishchev Papyrus, again after its first owner). All the documents date from the Middle Kingdom: that is, the first centuries of the second millennium BCE (about 2000–1600 BCE).

The Rhind Papyrus is particularly interesting because it contains a foreword that allows us to know the date when it was drafted (in approximately

---

[11] Herodotus, *Histories,* vol. II, 109, BUR, Milan, 2009, p. 441.

[12] Alice Cartocci, *La matematica degli Egizi. I papiri matematici del Medio Regno,* Florence University Press, Florence, 2007, pp. 8–12. Other fragments of mathematical papyri such as the Berlin Papyrus and the Kahun Papyrus do not add any substantial amount of knowledge.

the mid-17th century BCE) and the name of the scribe, Ahmose, who prepared it by copying it from older writings. The foreword is very ambitious compared to its purposes of mathematical calculation. It reads, "The proper way of entering nature, knowing everything that exists, every mystery, every secret." In actual fact, as in the case of the Babylonian tablets, the objective of the problems in the papyrus and the other available documents was essentially didactic and far more pragmatic. The problems focus on issues pertaining to daily life: the distribution of bread and food, calculations of fodder for livestock, the construction of granaries, and so on.

The following are some of the practical problems that recur frequently:

- calculate the area of a plot of land (rectangular, triangular, or trapezoidal) given the sides, and conversely, calculate one of the sides knowing the area and another side;
- calculate the volume of a granary in the shape of a parallelepiped or cylinder;
- calculate the distribution of loaves of bread among a number of men by imposing a given ratio between the first and the second, another ratio between the second and the third, and so on;
- knowing that with a certain amount q of wheat n loaves are produced, calculate how many loaves are produced if the amount of wheat used for each loaf is reduced to a given fraction (1/2, 1/3, or 1/10); similarly for brewing beer.

The notation used by the ancient Egyptians was additive decimal. The additive system facilitated addition and subtraction but made multiplication and division more complex, and so some simplifications were made. The first concerned multiplication, which was carried out by duplicating one of the operands and halving the other.

We can refer to a particularly simple case to illustrate the method. To multiply 27 by 8, we divide 8 by 2 (obtaining 4) and simultaneously multiply 27 by 2, obtaining 54. Then, we divide 4 by 2 (obtaining 2) and multiply 54 by 2, obtaining 108. Finally, we divide 2 by 2 (obtaining 1) and multiply 108 by 2, obtaining 216. Where one of the factors is not a power of 2 (if we need to multiply 27 by 11, for example), the procedure is a little more complex, as illustrated in the following. In order to better understand how an operation of multiplication by duplication and halving was carried out, we just need to note that it is the same as breaking one of the factors down into a sum of power of 2 (in the example, $11 = 1 + 2 + 8$) and multiplying the other factor by this sum by using the distributive property of multiplication: $27 \times 11 = 27 \times 1 + 27 \times 2 + 27 \times 8 = 297$.

This process is also illustrated by the following table:

Selected contribution	Breakdown of the second factor	Duplication of the first factor
v	1	27
v	2	54
	4	108
v	8	216
Total	11	297

In practice, actually, the algorithm proceeded faster by breaking down one of the factors not only into powers of 2 but also into multiples of 10. The scribe then added up the contributions needed to obtain the result. For example, to multiply 38 by 45 ($= 1 + 4 + 40$), one could proceed as illustrated by the following table.[13]

Selected contribution	Breakdown of the second factor	Multiplication of the first factor
√	1	38
	10	380
	20	760
√	40	1.520
	2	76
√	4	152
Total	45	1.710

[13] Cartocci, *op cit.*, p. 34.

Another way to simplify multiplication and division operations was to use only unitary fractions, that is, of the type $1/n$ (with the only exception of the fraction 2/3 also called *large half*, while 1/3 was called *small half*). Fractions with different numerators were reduced to unitary fractions by means of suitable conversion tables. For example, 3/10 = 1/5 + 1/10.

To calculate division between two numbers, one proceeded, in a way similar to multiplication, by duplicating the divisor as far as possible and then multiplying the divisor by fractions until the dividend was exactly reached.

Knowledge of algebraic calculus methods included the solution of first-degree equations (e.g., simple equations of the type $x + x/n = m$) and special cases of second-degree equations, which were dealt with by the false position method, a method widely used until the medieval time.[14]

An example of the application of the false position method to the case of first-degree equations contained in the Rhind Papyrus is as follows.[15] Suppose we have to solve the following problem:

A pile and its fourth part added together become 15.

It is easy for us to calculate the answer today by solving the simple first-degree equation $x + 1/4 x = 15$: The answer is 12. With the false position method, one proceeds in the manner described in the papyrus: We assume the (false) value 4 for x, then we add its fourth part — that is, 1 — and obtain 5. Then, we relate 5 to 15 and obtain a scaling factor of 3, and finally we multiply the initial assumption 4 by the scaling factor we have identified, from which we obtain the result 12.

Another aspect of arithmetic that is dealt with in the Rhind Papyrus is the computation of progressions.[16] It contains two problems on the computation of arithmetic progressions and one on the computation of a geometric progression. The following example is interesting because it

---

[14] Carl B. Boyer, *Storia della Matematica*, Mondadori, Milano, 1982, p. 19.

[15] Raffaella Franci (ed.), *Alcuino di York. Giochi matematici alla corte di Carlomagno*, ETS, Pisa, 2016, p. 16.

[16] Cartocci, *op cit.*, pp. 86–88.

corresponds to a nursery rhyme and is also found in medieval texts (in Leonardo Pisano's *Liber abaci*, for example):

> We have 7 houses, there are 7 cats in each house, 7 mice for each cat, 7 spelt plants for each mouse, and 7 *hekat*[17] for each plant; how many houses, cats, mice, spelt plants, and *hekat* are there in total?

The solution to the problem is not provided in the papyrus but is simply illustrated using the following calculations (we should recall that the sum of a geometric progression of the type $1 + r + \cdots + r^{N-1}$ is $(r^N - 1)/(r - 1)$):

$$s = 7 + 49 + 343 + 2{,}401 + 16{,}807 = 7 \times (1 + 7 + 49 + 343 + 2{,}401)$$
$$= 7 \times (7^5 - 1)/(7 - 1) = 7 \times (16{,}806/6) = (1 + 2 + 4) \times 2{,}801 = 19{,}607.$$

With regard to geometry, the area of mathematics the Egyptians are credited with inventing, we can confirm from the available mathematical records that they were able to calculate the areas and volumes of various geometric figures. To calculate the areas of circles and volumes of spheres (or hemispheres, such as a basket), the calculated value of π was very accurate (just slightly less than it was for the Babylonians): $256/81 = 3.16$.

One very interesting thing in the Moscow Papyrus is a calculation of the volume of a truncated pyramid using the formula $V = 1/3h\ (a^2 + ab + b^2)$, where a and b are the lengths of the larger and smaller sides, respectively, and h is the height of the truncated pyramid.

Finally, an observation should be made about the presentation style of the algorithms. Let us look at a simple example: the calculation of the area of a triangular piece of land taken from the Rhind Papyrus.

> If you are told: a triangle has a height of 10 *khet*[18] and its base is 4 *khet*.
> What is its area? Proceed as you must.
> Calculate half of 4, which is 2.
> Multiply 10 by 2. This is its area.
> Its area is 20.

---

[17] *Hekat*: a unit of volume measurement corresponding to about 4.8 liters.
[18] *Khet*: a unit of length corresponding to 100 cubits, or about 52 meters.

As we see, as in the case of Babylonian mathematics, Egyptian scribes presented algorithms by using numerical examples with a detailed description of the computational steps that were configured exactly like an algorithm and could therefore easily be applied to other numerical cases.[19] The Egyptians also used the structure of a description of the procedure to be followed, which began by indicating the problem to be solved and the data to be used and continued with the formula "Proceed as you must", followed by a predetermined structure that characterized the text as a sequence of instructions given to the student.

China

The dating of early Chinese mathematical documents is much less clear than it is for Mesopotamian and Egyptian documents, for various reasons. One of the oldest texts, the *Chou pei suan ching* (On the Gnomon and Circular Orbits of the Sky), is dated by some historians at 1200 BCE, while others attribute it to the early 3rd century BCE.[20] It takes the form of a dialogue between a prince and his minister. 246 problems are presented, mainly on the subject of astronomical calculations, geometrical and perspective issues (the calculation of shadows), and land surveying problems (which, as we have seen, were typical of agricultural societies based on the channeling of water and the control of river flooding). The introduction contains observations on the properties of right triangles (including a simple demonstration of Pythagoras' theorem) and calculations using fractions.

Among the oldest texts, the *Chiu chang suan shu* (The Nine Chapters on Mathematical Art) is also particularly interesting. It is believed to have been compiled in the 2nd century BCE, although it contains problems and results written down by several generations of scholars over the course of several centuries (probably from the 10th century BCE) and has come down to us through the works of other scholars who enriched it further up

---

[19] Jens Høyrup, *The Algorithm Concept — Tool for Historiographic Interpretation or Red Herring?* CiE 2008, Logic and Theory of Algorithms, LNCS 5028, Springer, Berlin, 2008, pp. 261–272.

[20] Boyer, *op cit.*, p. 229.

to the 13th century of our era. What makes this work particularly interesting from both an algorithmic and a historical point of view is that it is devoted to solving problems of practical interest, and thus allows us to understand the everyday needs that drove the evolution of mathematical computation. This text also presents 246 problems divided into nine chapters, and as in the Babylonian and Egyptian cases, it reflects the typical needs of a society based on agriculture and animal husbandry but with an advanced level of state organization.

The following are some of the problems considered in the text:

- exchange of products (millet, rice, etc.) with a calculation of proportions and various exchange rates;
- distribution of products and money among several people with various proportionality criteria;
- calculation of land yield and animal breeding and sale;
- calculation of taxes according to different taxation criteria;
- calculation of areas of fields of various shapes: rectangular, triangular, trapezoidal, and circular;
- calculation of volumes related to constructions;
- reverse problems from the previous ones (calculation of sides, diameters, etc., given areas of surfaces and volumes).

The following example, which is taken from Chapter 2, illustrates the kinds of topics discussed in the text:

> 2 1/2 *picul* of rice are bought for 3/7 of a silver *taiel*; how many *picul* can be bought with 9 *taiel*?

Here, *picul* and *taiel* are evidently measures of volume and cost, respectively. The solution, which is obtained by calculating a simple proportion, is 52.5 *picul*.

Another paradigmatic problem (which also appears in other mathematical texts up to the medieval period) is that of the broken bamboo cane. Given the height of a bamboo cane and the point at which it has broken, the problem is to calculate the distance between the point where the top of the cane touches the ground and the point where its base is embedded in the ground. The exercise involves a knowledge of Pythagoras' theorem and an ability to calculate square roots.

Although the text makes no use of algebraic notation and does not include any demonstrations, it should be noted that the mathematical notions required by problem-solving are nevertheless very broad. In addition to integer and fractional arithmetic, they include calculating the greatest common divisor, calculating proportions, extracting square and cube roots, calculating progressions, solving some types of first- and second-degree equations by using the false position method, and solving systems of linear equations by applying methods that are not dissimilar to Gaussian elimination. As for geometry, it is interesting to note that one of the chapters, the ninth, is devoted to right triangles and begins with a version of Pythagoras' theorem.

Chinese mathematics developed further in the early centuries of our era, seemingly quite independently of external input from India, Greece, or the Middle East. According to the mathematical history scholar Raffaella Franci, two texts in particular, *Master Sun's Handbook of Mathematics* and *Zhang Qiujian's Handbook of Mathematics*, from the 4th and 5th centuries CE, contain interesting collections of problems that we might call exercises in mathematics applied to everyday life with aspects of recreational mathematics.

For example, the second of these texts contains the first known version of the Hundred Birds Problem, which we find again later in Arabic mathematical texts.[21] One of the most noteworthy results obtained during this period is the precision achieved in the calculation of the value of π – 3.14159 — which was obtained by Liu Hui in the third century CE using a 3,072-sided regular polygon.

Finally, while noting the quality and breadth of Chinese mathematical knowledge and computational skills, we must emphasize that in the early 17th century, the Italian Jesuit Matteo Ricci found a significant flaw in Chinese mathematics from a logical-deductive standpoint. He therefore thought it appropriate to teach the Chinese the formal methods of European mathematics, with particular reference to the mathematics developed by the Greeks. His main contribution in this regard was a

---

[21] Franci (ed.), *op cit.*, p. 18.

translation into Chinese of the first six books of Euclid's *Elements*, which he then published. In the preface to his work, he underlined that "without a good foundation it is difficult to successfully develop anything solid, which is why even the most informed scholars are unable to prove the assumptions behind their conclusions."[22]

Greece and Rome

The beginning of Greek mathematics is traditionally dated to the 6th century BCE, the century of Thales (c. 624–548 BCE) and Pythagoras (c. 580–500 BCE), although the absence of documentary evidence leaves a shadow of doubt about their lives and activities. It is fairly certain, however (once again according to tradition), that both visited Mesopotamia and Egypt (and Pythagoras may also have traveled to India) and that these contacts enabled the studies of astronomy and mathematics carried out in these countries in previous centuries to penetrate Greece. What is certain is that from the 6th century BCE onward the development of mathematics in Greece took on an intensity and a dimension that had no equal in the ancient world. In addition to the unquestionably exceptional quantitative aspect, the development of Greek mathematics had two important features: one methodological, namely the axiomatic deductive approach that underlay the procedures that were applied, and the other an interest in the abstract properties of the mathematical entities being studied (especially numerical and geometric entities).[23]

These attributes give Greek mathematics a very different perspective from the one seen in the case of Mesopotamia, Egypt, and China. Given the documentation that has been passed down to us from these ancient cultures on the mathematical operations necessary to perform everyday activities, it seems strange that a civilization like that of the Greeks, which could boast some of the most excellent mathematical minds, has not handed down to us any works devoted to the applied and practical aspect

---

[22] Jonathan D. Spence, *Il palazzo della memoria di Matteo Ricci*, Adelphi, Milan, 2010, p. 89.

[23] Ahmed Djebbar, *Storia della scienza araba. Il patrimonio intellettuale dell'Islam*, Raffaello Cortina Editore, Milan, 2002, p. 180.

of calculus. According to the historian of mathematics Carl Boyer, "The Hellenes had a reputation for being shrewd merchants and traders, and there must have existed a lower level of arithmetic or calculus that satisfied the needs of the great majority of Greek citizens," but "this type of operation would not have been worthy of any consideration by the philosophers."[24]

The idea that the Greeks were not interested in the applications of mathematics has been supported over time by the reflections of historians of Greece and Rome as well as by various anecdotal observations. It is said, for example, that when a pupil asked Euclid why the study of geometry was useful, Euclid ordered a slave to give him a coin because, he said, "He needs to make a gain out of what he learns." It is even said that Archimedes, a mathematician who made such fundamental contributions to applied physics, despised the practical results he achieved.

According to the historian of science Lucio Russo, on the other hand, only the simultaneous presence of studies of an applied nature (in the fields of mechanics, optics, and astronomy) justifies the development of axiomatic deductive studies by the Greek mathematical school.[25] In other words, speculative studies intended to attain abstract, formal results alone could not have existed if engineering components aimed at resolving problems of a practical nature that had the ability to exploit the results obtained by theoretical mathematicians had not been a part of the social fabric.

In fact, there are manifold applications of the mathematical results obtained by the Greeks in the fields of astronomy (for instance, the measurement of the distance of the Earth from the Moon and the Sun by Aristarchus or Ptolemy's contributions in the Almagest), geodesy (for example, the measurement of the Earth's meridian by Eratosthenes), and various other matters that we might define as mathematical physics, such as those studied by Euclid (relating to optics) and Archimedes (relating to hydraulics). On the other hand, it is a well-known fact that the motives that were the driving force behind the development of Greek mathematics were more about the search for the logical perfection of mathematical

---

[24] Boyer, *op cit.*, p. 68.
[25] Lucio Russo, *La rivoluzione dimenticata*, Feltrinelli, Milan, 1998, pp. 111–116.

assertions than their interpretation for practical purposes. The link between mathematics and philosophy that was a feature of Greek mathematical studies is one — but not the only — explanation that enables us to understand the originality and, one might say, uniqueness of Greek mathematics. Indeed, it is also possible to establish a connection between the abstract, logico-deductive approach of Greek mathematics and the development of dialectics imposed by the embryonic forms of democratic life in Greek society and the resulting need to have a rational basis for argumentation and persuade an adversary to accept a particular conclusion that has been deduced logically from the premises.[26]

The fact remains that the Greeks not only distinguished between the theory and application of mathematics but also believed that there was a difference between a numerical calculation carried out with traditional — or one might say mechanical — procedures, which was given the name "logistics",[27] and the study of the properties of numbers. Logistical calculations could also be carried out by slaves, while the development of a theory was the result of a more elevated cultural approach.

In this context, geometry had a privileged role for the Greeks. It was a terrain that lent itself to axiomatic deductive reasoning, and so we should not be surprised that the Greeks replaced the algebra developed by the Mesopotamians, which we might term "arithmetic", with a "geometric" algebra, the results of which would be obtained by establishing a correspondence between numbers and numerical properties on the one hand and geometric entities on the other.[28] This approach was used, for example, to calculate the powers of a binomial or the solution to second-degree equations.

In the most important Greek mathematical text we have, Euclid's *Elements*, the first six books are devoted to geometry. From an algorithmic point of view, it is interesting to note that even one of Euclid's most classic and scholastic contributions, the algorithm for the greatest common divisor, uses a geometric approach in which numbers are replaced by lengths

---

[26] Boyer, *op cit.*, p. 92.

[27] *Ibid.*, pp. 146–147.

[28] As we will see in the next chapter, this approach was also widely used in the works of Arab mathematicians between the 8th and 10th centuries.

of segments. There are two especially interesting and innovative aspects of the algorithm for the greatest common divisor in Euclid's *Elements* that are fundamentally related not to the substance of the algorithm but to the form in which it is presented. The first is that in this case, unlike in the description of Babylonian and Egyptian algorithms, the data are not numerical but symbolic. The presentation of the algorithm begins thus: "Suppose that two numbers [that is, two segments of length] AB and ΓΔ are given that are not prime to each other, and ΓΔ is the smaller." It then goes on to show that a segment ΓZ whose length corresponds to the greatest common divisor of the length of AB and the length of ΓΔ must be determined. The second aspect is that in a certain sense the algorithm also contains proof of its own correctness, and ends with the following statement: "What was to be proved."

If we want to find a contribution in the field of Greek mathematics that offers a departure from the traditional abstract logical-deductive approach and expresses considerable algorithmic and computational interest — albeit with no reference to practical applications — we have to wait for a much later historical period. It was in the 3rd century of our era that Diophantus, a mathematician who lived in Alexandria, Egypt, compiled the thirteen volumes of his most important work, the *Arithmetica*, which is devoted to solving approximately 150 problems that are presented with specific numerical values and for which precise complete solutions are sought. These problems basically correspond to equations, or systems of equations, in one or more unknowns, determinate or indeterminate (which are still called Diophantine equations today). Of the thirteen volumes that originally made up the work, six are available in the original Greek and four in an Arabic translation. Overall, Diophantus' work can be viewed as a collection of applied algebra problems, and as such, it is closer to Mesopotamian mathematical texts (which were probably not unknown to Diophantus) than to traditional Greek mathematics.[29]

The first types of problems addressed by Diophantus can be traced back to solving first-degree equations, such as the heap problems in the

---

[29] Boyer, *op cit.*, p. 214. According to Lucio Russo, the discovery of cuneiform texts "drastically reduced Diophantus's originality, showing that the methods he described had been in use in Mesopotamia for a long time."

Rhind Papyrus that we have seen being solved using the false position method. A singular example of this kind of problem (which is not in the text but is inscribed on the author's tomb) is the epitaph that Diophantus himself wanted on his tomb and which poses a riddle for passers-by:

This tomb holds Diophantus. Ah, how great a marvel!
The tomb tells scientifically the measure of his life.
God granted him to be a boy for the sixth part of his life,
and adding a twelfth part to this, he clothed his cheeks with down;
He lit him the light of wedlock after a seventh part,
and five years after his marriage. He granted him a son.
Alas! late-born wretched child;
after attaining the measure of half his father's life, chill Fate took him.
After consoling his grief by this science of numbers for four years he ended his life.[30]

The answer to the riddle is found by solving a simple first-degree equation, $x/6 + x/12 + x/7 + 5 + x/2 + 4 = x$, from which it can easily be calculated that Diophantus died at the age of 84.

However, Diophantus' work also contains problems of greater complexity that require solving second-degree equations.

A classic example (Problem 27 in Book I) is finding the length of the two sides of a rectangle, x and y, so that their sum equals 20 and the area of the rectangle equals 96: that is, $x + y = 20$, where $xy = 96$. The solution to this problem was already present on Babylonian tablets used for the education of scribes, and Diophantus solved it using a similar method:

We divide the sum of the two sides by 2, and obtain 10; the two sides will therefore be length $x = 10 - z$ and $y = 10 + z$ long.
When we solve the equation $(10 - z)(10 + z) = 96$, we obtain the value 2 for z. The lengths of the two sides are therefore 8 and 12.

---

[30]This epitaph is among the 45 riddles in the form of epigrams corresponding to first-degree equations collected by the Greek collector of epigrams Metrodorus (6th century CE) and later included in the 11th-century codex discovered in Heidelberg known as the Palatine Anthology.

Other examples of problems include the following: Find two numbers where their sum and the sum of their squares are equal to two given integers, and look for solutions to the indeterminate equation $x^2 = a + by^2$.

A final particularly important example is the search for integer solutions to the equation $x^2 + y^2 = z^2$. This equation admits an infinite number of solutions that are commonly known to consist of the so-called *Pythagorean triples* (3, 4, 5 or 5, 12, 13, etc.),[31] which had also been known by the Mesopotamians two thousand years earlier.

One limitation of Diophantus' work — which now seems to us to be decidedly unsatisfactory — is that for equations higher than the first degree, only one solution is given — the smallest — if there are two positive solutions, or possibly the only positive solution, which even more seriously is also the case with indeterminate equations, which can have many numbers or an infinite number of solutions. Despite these limitations, Diophantus is considered, not without justification, to be the father of algebra. As we have said, although he refers to specific instances of problems and thus does not have the level of abstraction that can be found in other Greek texts, Diophantus shows a solution by using what for us are now classic algebraic methods (and not, for example, geometric methods). What is more, his style of presenting problems and solution methods also marked an advance over the earlier so-called rhetorical style, which consisted in describing unknowns, operations, and numbers textually. In Arithmetica, "there is a systematic use of abbreviations to indicate powers of numbers and to express relations and operations"[32] in accordance with a style known as syncopated. In a sense, therefore, Diophantus' approach lies halfway between the rhetorical style and the one we use today, in which operators and operands are represented in a purely symbolic form.

---

[31] Andrew Wiles showed in 1995 that there are no integer solutions for equations like $x^n + y^n = z^n$ with $n > 2$. This question is known as Fermat's Last Theorem, and the French mathematician Pierre de Fermat was specifically inspired to address it by reading Arithmetica. In the margin of Diophantus' text, Fermat noted that he had found a "marvellous demonstration of this theorem" (Amir D. Aczel, *Fermat's Last Theorem*, Delta, New York, NY, 1996, p. 9; see also Pierre de Fermat, *Osservazioni su Diofanto*, Bollati Boringhieri, Turin, 2006). In fact, however, the only known demonstration of the theorem was provided 350 years later by the English mathematician Andrew Wiles by the use of refined analytical and geometrical techniques.

[32] Boyer, *op cit.*, p. 212.

As far as the Latin world is concerned, it is a well-known fact that the Romans did not produce any particularly significant mathematical works, and so it must also be acknowledged that there are only very few mathematical texts of a practical nature that relate to solutions for problems of everyday life. One of the few contributions we have is the *Corpus Agrimensorum Romanorum*, which is dedicated to measuring land for the purposes of buying and selling it or taxation or calculations of yield. It is a collection of texts by various authors from the imperial period (1st century BCE-4th century CE), including Frontinus, Siculus Flaccus, and Agenius Urbicus, which became the basis for the texts on geometry that were used until the early medieval period.

The limited documents on applied mathematics in the Greek and Roman worlds does not mean that calculus was not used intensively in everyday life for a wide variety of applications. Clearly, the use of additive notation to represent numbers, as used by the Romans, did not make calculations easier. For example, in order to perform multiplications, it was necessary to use the method we described earlier of multiplying one of the two factors by 2 and dividing the other factor by 2. With regard to methods of calculation, based on what we read in various Latin texts, we know that the Romans used a form of digital calculus that allowed numbers from 1 to 9,999 to be represented by the position of the fingers of both hands. The fingers of the left hand (outstretched or bent) represented the numbers from 1 to 9 and tens from 10 to 90, while hundreds and thousands were represented by the fingers of the right hand.[33] This method, which had ancient origins and was also used in the Middle East, had the advantage of being simple to use and having a secret and mysterious air. As we will see, it continued to be used over the centuries.

However, the practical tool that was primarily used by the Romans to make calculations was the abacus.[34] Skilled use of the device (particularly

---

[33] Georges Ifrah, *Histoire universelle des chiffres*, vol. I, Laffont, Paris, 1994, pp. 138–139.

[34] The term *abacus* comes from the Greek word αβαξ, meaning "tablet" and earlier from the Semitic word *abq*, meaning "powder", because the earliest versions of the instrument consisted simply of a powder-covered board on which numerical symbols and geometric figures were traced. It seems that the earliest uses of the abacus date back to the Babylonians (around the 4th century BCE).

the "portable" version) allowed merchants, engineers, tax collectors, and others to perform computations anywhere (as was customary in China, Japan, Russia, and Turkey even until recently). The abacus used by the Romans (of which there were several variations) generally consisted of a tablet of wood with a series of grooves, at the top of which characters corresponding to the values of the Roman numbering system were engraved from right to left: I for units, X for tens, C for hundreds, and M for thousands. Pebbles (*calculi*) indicating the number of units, tens, hundreds, etc., that were to be represented were placed in these grooves.[35] The absence of pebbles in a column indicated the absence of that power of 10 in the number represented, and therefore in fact, the abacus was already an embryonic form of the positional system and presumed the existence of zero. It also had the virtue of being adaptable to different numerical systems. Performing arithmetic addition or subtraction operations was naturally quite simple, but multiplication and division were much more difficult.

It was not until the 13th century, when the positional numerical system that had originated in India became established thanks to works by Arab mathematicians, that the use of the abacus began to be supplanted by numerical algorithms, although it was still used until the 17th century. A famous illustration in the *Margarita Philosophica* (an encyclopedic work by the German humanist Gregor Reisch first published in 1503) depicts a comparison between an algorithmist (Boethius, who was credited with the spread of Indian numerals[36] and related computational algorithms in Europe) and an abacist (Pythagoras, who was credited with the invention of the abacus), both of whom are engaged in performing a computation (Figure 2).

The supremacy of the algorithmic method of calculation is indicated in the image by the fact that while the algorithmist has completed his calculation and has laid his stylus on the table, the abacist is still calculating. Moreover, a feminine depiction of Arithmetica is portrayed standing behind them with the Arabic-Indian numerals of the two sequences 1-2-4-8 and 1-3-9-27 (the first values of the powers of 2 and 3) embroidered

---

[35] As we know, the word *calculus* comes precisely from the Latin word *calculus*: It refers to the pebbles that were used in an abacus.

[36] Whether this claim is true will be discussed in Chapter 3.

Figure 2.   A challenge between an abacist (Pythagoras) and an algorithmist (Boethius). Behind them stands Arithmetica (Gregor Reisch, *Margarita Philosophica*, 1503). (Private Collection, © Giancarlo Costa/Bridgeman Images).

on her dress. In this way, she seems to enshrine her preference for calculation methods that use Arabic numerals over the abacus.[37]

---

[37] Nadia Ambrosetti, *L'eredità arabo-islamica nelle scienze e nelle arti del calcolo dell'Europa medievale*, LED, Milan, 2008, p. 13. See also Luigi Laura, *Breve ed universale storia degli algoritmi*, Luiss University Press, Rome, 2019, pp. 29–30.

India

Indian mathematics played a very important role in the overall development of mathematics globally. Although this role has perhaps been less significant than has sometimes been claimed,[38] and above all has lacked continuity, there is no doubt that Indian mathematicians were instrumental in the adoption of the base-10 positional numbering system and more generally from the point of view of numerical calculation techniques.

In the case of India, the earliest body of mathematical knowledge on which we have information had its origins in computational needs arising from everyday life. It is strikingly analogous to the texts by Egyptian rope-stretchers, which, as we have seen, are linked to the origins of the development of geometry, and describe the use of ropes to measure land and build temples and altars. These texts are known as the *Sulvasutras* (literally, rules of the ropes) and they date from around 800 BCE, although later versions, all in verse, were written up to 200 CE. The *Sulvasutras* include rules on the construction of right angles using ropes of lengths corresponding to Pythagorean triples (for example, 3, 4, 5 or 5, 12, 13). They also deal with other geometric problems, such as the construction of squares of an equal area to a given rectangle. This suggests that certain results of Babylonian and Greek mathematics were known in India.

The next important Indian mathematical work also draws its motivation from applications. This is the *Siddhanta* (System of Astronomy), which dates from the 4th–6th centuries CE. Five different versions of this work are known,[39] one of which, the *Surya Siddhanta* (System of the Sun), may be the only one we have in its entirety. The most important contributions of the *Siddhanta* include a definition of the concept of the sine of an angle that is substantially similar to the definition we give it today and the calculation of sine tables for 24 evenly spaced angles between 0 and 90 degrees.

---

[38] Vedic mathematical texts such as *Vedic Mathematics* by Jagadguru Shankaracharya Swami Bharatikrishna Tirtha (Motilal Banarsidass Publishing House, Delhi, Varanasi, and Patna, 1989) claimed that all mathematical knowledge was encapsulated in sixteen sutras written in Sanskrit in very ancient times. See also Narinder Puri, *An Overview of Vedic Mathematics*, Rajasthan University, Jaipur, 1988.

[39] Boyer, *op cit.*, pp. 244–245.

In general, both the *Siddhanta* and the later works by Aryabhata and Brahmagupta (6th–7th centuries CE) demonstrate some important indebtedness to Egyptian, Babylonian, and Greek mathematics, but they also have some original aspects, such as the identification of negative solutions of second-degree equations. It should be emphasized, however — and this is the most important aspect with respect to the subject matter of this book — that the main contribution of India to the evolution of mathematics was the introduction of the decimal positional system based on the nine symbols corresponding to the digits between 1 and 9 and the use of zero, which made it possible to introduce efficient algorithms for arithmetic operations. In the Indian numbering system, in fact, zero is not just a placeholder required to mark the absence of a power of the base, as it was in Mesopotamia; it becomes a constituent element of the calculation. Brahmagupta in particular makes an important contribution from this point of view by presenting a comprehensive treatment of arithmetic operations between positive numbers, negative numbers, and zero.

The Indian numbering system spread rapidly, first to the Middle East, then throughout the area under Islamic domination, and finally to Europe. According to some texts, the system was introduced into Baghdad at the end of the 8th century by a Hindu scholar, Kankah, who is also said to have taken a Persian version of the *Siddhanta* to Baghdad that was later translated into Arabic. One of the most valuable sources in this regard is the Arab historian Ibn Khaldun, who was born in Tunis in 1332 and died in Cairo in 1406. In his text *Muqaddimah* (Prolegomena), he wrote that the Arabs inherited science from India together with their numerals and methods of calculation when a group of wise men from India presented themselves to Caliph al-Mansur in the year 156 of the Hegira (773 CE).[40] According to other sources, "Indian numerals" penetrated the Arab world as early as the mid-7th century. In fact, the writings of a Syrian bishop, Severus Sebokht, which date back to 662 CE, mention the "precious methods of calculation of the Indians that were carried out by means of nine signs."[41]

---

[40] Ifrah, *op cit.*, vol. II, p. 239.
[41] Boyer, *op cit.*, p. 250.

Chapter 2

Algorithms in Daily Life at the Time of the House of Wisdom

The House of Wisdom

As mentioned in the previous chapter, the first territories to which the Indian positional numbering system and related algorithms for arithmetic operations spread were those under Islamic rule. It was from those territories that the computing methods that we can call Indo-Arabic later spread to the European West as well.

In the history of the sciences and especially of mathematics and computation, the Arabs played a particularly important role. More generally, from the 8th until the 12th century, a period known as the "golden age of Arab science", the world center of scientific studies — especially in the fields of astronomy, mathematics, chemistry, and medicine — was established in a vast area extending from present-day Iraq to Syria, Persia, Egypt, the Maghreb, and Spain, westward and eastward, to the Central Asian territories of Khoresme (or Khorasmia) and Transoxiana, which are now part of Uzbekistan. All these regions were unified by the Islamic religion and the use of the Arabic language.

In fact, during the 7th century, due to the early establishment of Islam in the Middle East, the region was not particularly open to the scientific world but saw an obscurantist phase in which radical and irrational religious drives prevailed. According to tradition, Omar, the second caliph, who ruled from 634 to 644, had a large number of books — that had been taken out of Iran at the time of the conquest in 637 — destroyed, declaring that if they contained truths, those truths would be in any case superseded by Allah's revelation in the Qur'an, and if they contained falsehoods they should still be destroyed. The same episode is reported by the historian al-Baghdadi (12th century) with reference to Omar's decision to destroy the books of the library of Alexandria in 642 when the city was conquered.[1] However, this period had a limited duration. Already under the Umayyad Caliphate, beginning in 661, schools were established in Syria where numerous Greek philosophical and scientific works were translated. There are reports, for example, that the aforementioned Syriac bishop Sebokht translated and commented on the works of Aristotle and, based on Greek sources, compiled texts on astronomy and geography.[2]

Interest in Greek philosophical and scientific sources was further strengthened under the Abbasid Caliphate, beginning in 750. A few years after the Abbasids established themselves over the Umayyads, the second Abbasid caliph, al-Mansur (who ruled in the years 754–775), decided to found a new city, Baghdad, on the banks of the Euphrates River. According to reports by a 10th-century historian, al-Tabari, the site was chosen directly by al-Mansur and the foundation was established on July 30, 762 on the instructions of three astrologers who had been consulted for the occasion.[3] After the capital of the caliphate was moved to Baghdad, which was the second capital after Kufa, the city quickly became a major center not only of trade but also of culture. The Abbasids fostered the development of a tolerant, multicultural society in which Arabs, Jews, Christians, and Zoroastrian Persians coexisted peacefully and contributed to

---

[1] Boyer, *Storia della Matematica*, p. 264. There have been several occasions when books from the library of Alexandria were destroyed during events of war (as in the case of the war between the Roman emperor Aurelian and the Palmyra queen Zenobia in the 3rd century) or deliberately, in opposition to "pagan" literature by Muslims (as in the above-mentioned case of Caliph Omar) or Christians (as at the time of the edict of Theodosius, in the 4th century). Even the attitude of some Christian thinkers, such as Tertullian in the 2nd century, had not been dissimilar to that of Muslims. Indeed, they too believed that scientific research had become useless after the spread of the Gospel.

[2] Adolf Youschkevitch, *Les mathématiques arabes*, Vrin, Paris, 1976, pp. 3–4.

[3] Jim Al-Khalili, *La casa della saggezza*, Bollati Boringhieri, Torino, 2010, p. 60.

prosperity. It was in this context that the golden age of Arab science was born and developed.

The prime mover behind the cultural and scientific development of those years — a development that grew and continued from the 8th to the 12th century, albeit with ups and downs — was the proliferation of translations of Persian, Indian, and Greek philosophical and scientific texts into Arabic. As mentioned earlier, these were the years when the Indian numeration system and its methods of calculation were imported from the Arabs and when the astronomical work *Siddhanta* was translated into Arabic in Baghdad (with the title *Sindhind*). Shortly afterward, the astrological work Tetrabiblos by Ptolemy was also translated into Arabic from Greek.[4]

After al-Mansur, Caliph Harun ar-Rashid (786–809) — who is best known along with Shahrazad as the main character in the *Arabian Nights* — continued to fuel the cultural growth of the capital of the Islamic empire. In particular, ar-Rashid, who was himself a scholar and poet, founded an important library in Baghdad filled with volumes, mainly from Constantinople. Among the most important Greek works translated at that time were Euclid's *Elements* and Ptolemy's *Almagest*; both works were translated by al-Hajjaj ibn Yusuf ibn Matar, but they were later retranslated several times as knowledge of the topics covered deepened. Various Persian families who had supported the Abbasid assertion over the Umayyads then enjoyed great prestige in Harun ar-Rashid's court and held an important role. Until their presence became too intrusive so as to cause ar-Rashid's hostility, it influenced the translation movement and increased the contribution of Persian science and culture in the Arab world.

Several years of great instability followed ar-Rashid's death. Ar-Rashid's designated successor as caliph was his son al-Amin, who was soon opposed, however, by his other son, al-Ma'mun, who enjoyed the support of the empire's eastern territories. With the help of Persian troops, al-Ma'mun gained the upper hand and became caliph in 813, a role he retained until his death in 833.

The contribution of al-Ma'mun to the development of science in the Islamic world has been fundamental. One legend has it that when he

---

[4] Boyer, *op cit.*, p. 265.

resided in Merv as governor of Khorasan, he saw Aristotle in a dream, and from then on, a real obsession with Greek culture and philosophy developed in him. The reality is that he is credited with sending emissaries to Constantinople and other cities to collect Greek books and scientific documents and further credited with fostering the translation of Greek, Persian, and Indian texts into Arabic, making Arabic the *lingua franca* of science in the last centuries of the first millennium. This cultural development was nurtured by both the extremely good economic period and the fact that the early Abbasid caliphs were influenced by the liberal, rationalist Islamic current of the Mutazilites. According to the Mutazilite theological school, it is intellect that leads man to morality and knowledge of God. The Mutazilites were proponents of free will and the primacy of philosophical reasoning over revealed religious dogma.[5]

During his caliphate, in around 820, al-Ma'mun founded a kind of academy in Baghdad called *Bayt al-Hikma* (House of Wisdom) that housed a library (*khizanat al-hikma*) and an astronomical observatory. It is believed that by the mid-9th century, the library of the House of Wisdom had become the largest repository of books in the entire world.[6] Scholars from many regions of the Islamic world were invited to the academy by al-Ma'mun: mathematicians, astronomers, geographers, and natural science scholars.

Among the most important figures who worked at that institution in the early 9th century, Hunayn ibn Ishaq, the philosopher al-Kindi, and the biologist al-Jahith are worth mentioning. The former, who was appointed by al-Ma'mun as head of the translators of the House of Wisdom, translated works by Plato and especially Galen (*On the Anatomy of Veins and Arteries, On the Anatomy of Nerves*, etc.) and also carried out research on the anatomy and physiology of the eye. Al-Kindi, besides being a philosopher and musicologist, was also a mathematician, especially an expert in cryptography. His works on philosophy, in which he tried to reconcile Aristotelian philosophy with the Islamic religion, were very important. Al-Jahith, who

---

[5] The Mutazilite school of theology was always opposed by the Orthodox currents that took over in the late 9th and 10th centuries and asserted the primacy of divine revelation. The Mutazilite school disappeared completely in the 13th century.

[6] Al-Khalili, *op cit.*, p. 106.

was probably acquainted with Aristotle's work *History of Animals*, in turn wrote a book on this subject in which he hypothesized the existence of common ancestors for species of animals with similar characteristics (e.g., dogs, wolves, and foxes) and examined the question of the inheritance of acquired behaviors. The theories of both al-Kindi and al-Jahith contributed to reinforcing al-Mamun's rationalist religious orientation.[7]

Finally, the role played at the House of Wisdom by the Banu Musa brothers should be mentioned. Taken under the protection of Caliph al-Ma'mun, the three brothers, Muhammad, Ahmed, and Hassan, devoted themselves to studies in mathematics, physics, music, and astronomy. Among the major works they wrote were the *Book on the Measurement of Plane and Spherical Figures*, the book on *Book on the First Motion of the Celestial Sphere*, and, particularly interestingly, the *Book of Ingenious Devices*, which contains plans for automatic machines such as a flute player and a water organ.

Al-Khwarizmi

Among the scholars who converged on Baghdad at the House of Wisdom, many came from Persia and neighboring regions of Transoxiana. Indeed, the support of the Persian and Khorasanian elements for the Abbasid establishment and the support offered to al-Ma'mun had resulted in a reinforcement of contact between the Arab world and the Persian and Central Asian worlds (Khorasan, Khorasmia, and Transoxiana), and it should be noted that a large part of Baghdad's population also came from these regions.[8]

From the city of Khiva, in Khorasmia, came what can rightly be considered the most important mathematician and astronomer who worked at the House of Wisdom in the 9th century and who was destined to leave his name in the history of mathematics and computer science: al-Khwarizmi.

---

[7] Al-Khalili, *op cit.*, pp. 110–111.
[8] Frederick S. Starr's text, *Lost Enlightenment: Central Asia's Golden Age from Arab Conquest to Tamerlane* (Princeton University Press, 2013), highlights the strong interconnections between the Central Asian world and the Arab world since the 8th century and the contribution that the culture of the Central Asian regions made to Arab science.

Figure 1.   Statue of al-Khwarizmi in front of the Faculty of Physics and Mathematics at Urgench University (Uzbekistan)

Abdallah Muhammad ibn Musa al-Khwarizmi al-Magusi was born in around 780 and died in 850 (Figure 1). His name indicates that he came from the city of Khiva in Khorasmia, and the appellative al-Magusi may indicate descent from a family of Zoroastrians, the religion of Zoroaster being widespread in that region. Information about his life is very sparse.[9] His works concern not only mathematics but also astronomy, geography, history, calendars, sundial construction, and the astrolabe. Unfortunately, many of his works have been lost and we have knowledge of them only thanks to the bibliographical lists compiled by some Arab authors.

The work on astronomy *Zij al-Sindhind* has come down to us only through a version re-elaborated in the 10th century by an astronomer from

---

[9] Statues dedicated to al-Khwarizmi, called al-Khorasmiy in Persia and Uzbekistan, can be found in the city of Khiva and in the capital of Khorasmia, Urgench. Unfortunately, the imposing statue that stood in Khiva was recently removed. The faculty of Physics and Mathematics in Urgench bears his name. The Uzbek state dedicated a postage stamp to him on the 1,200th anniversary of his birth. According to some historians, he was educated in Merv, Khorasan, the city where al-Ma'mun resided for a few years from 813 to 819, and did not move to Baghdad until he was in his forties, following the caliph; see Starr, *op cit.*, According to other scholars, al-Khwarizmi always lived in Baghdad and the name only indicates where his family came from; see Roshdi Rashed, *Al-Khwarizmi. The Beginning of Algebra*, Saqi, London, 2009, p. 6. Rashed himself argues that the attribution to al-Khwarizmi of the appellation al-Magusi may derive from a copying error by a copyist.

Cordova, al-Magriti, and was probably translated into Latin by Adelard of Bath. *Zij al-Sindhind* is based on the Indian work *Siddhanta*, as well as Greek and Persian sources, and contains 116 tables of astronomical data concerning the Sun, the Moon, and the five planets then known, as well as tables of the values of sines, cosines, and tangents.[10]

The geographical work *Kitab Surat al-Ard* (*Book of the Description of the Earth*) is another work of great importance; finished in 833, it has come down to us in Arabic and Latin. The full title is *Book of the Description of the Earth, with Its Cities, Mountains, Seas, All Islands and Rivers* written by Abu Jafar Muhammad ibn Musa al-Khwarizmi following the geographical treatise composed by Ptolemy. The work is actually based on Ptolemy's geography and is organized by climate bands (i.e., in relation to latitude), and each band is then presented according to longitude. Compared with Ptolemy, al-Khwarizmi does, however, provide some interesting innovations. In particular, he reduces the extent of the Mediterranean (bringing it back fairly correctly to 50 degrees longitude) and also represents the Atlantic and Indian Oceans as open sea areas and not enclosed by land.

Finally, the work on the Hebrew lunisolar calendar deals specifically with the so-called "Metonic cycle"[11], that is, the 19-year cycle after which the moon phases recur on the same days of the year, a cycle on which the determination of the day of the Passover is based. This work shows that al-Khwarizmi also had a thorough knowledge of Jewish culture.

Al-Khwarizmi wrote two fundamental works on mathematics. One, the original of which has been lost and is known only through Latin translations, concerns the Indian numbering system and related methods of calculating arithmetic operations. The other, of which the original is

---

[10] Whether the tangent tables are actually the work of al-Khwarizmi is not certain. They may have been inserted later. Indeed, it seems that the concept of tangents was introduced by the Persian mathematician Abu 'l-Wafa in the second half of the 10th century. See Youschkevitch, *op cit.*, p. 51.

[11] The 19-year lunar cycle is named the "Metonic cycle" after the Athenian astronomer Meton, who is said to have determined it in 432 BCE, but in fact it appears to have been known earlier in the Mesopotamian and Hebrew worlds, as far back as the sixth century BCE.

known, on the other hand, concerns the solution of first- and second-degree equations and is of particular importance because, for the first time, the term "algebra" is used in the title and it is therefore considered the beginning of that discipline.[12] In addition, what makes it particularly significant is that, with respect to the purpose of this book, al-Khwarizmi's mathematical works present content of both theoretical and practical interest, especially taking into account the needs of daily life.[13]

According to a bibliography compiled by al-Nadim[14] (a scholar and copyist working in Baghdad who died in 995), the first work may have had the title *Kitab al-jam' wa-al-tafriq al-hisab al-hindi* (*Book of Addition and Subtraction According to Indian Calculus*). As noted earlier, the original has been lost, and several Latin versions of this work are known. The following are the most important of them: a 13th-century manuscript preserved at Cambridge University called *De numero indorum* or *Dixit Algorizmi*, from the words with which it begins; a translation, probably by John of Seville in Toledo from the 12th century (Toledo was at that time a hotbed of Latin translations of Arabic scientific and philosophical works), entitled *Liber Algorismi de practica arismetricae*; and a translation edited from 1143 that, according to some people, was realized by Adelard of Bath, who was also working in Toledo, entitled *Liber Ysagogarum Alchorismi in artem astronomicam a magistro A. compositus*. The work is devoted to arithmetic operations according to the Indian numbering system. It consists of twelve chapters that deal gradually with

---

[12] According to some authors, the work by the Central Asian mathematician 'Abd al-Hamid Ibn Turk, which has the same title as al-Khwarizmi's work (*Kitab al-jabr wa-al-muqabala*) and contains some identical examples, was compiled shortly before and thus it should be recognized as the first algebra text. However, it should be emphasized that in any case, because of its completeness and clarity, al-Khwarizmi's work deserved to become a fundamental reference for centuries in Arab countries and in the West; see Boyer, *op cit.*, p. 273.

[13] Youschkevitch, *op cit.*, p. 15.

[14] The bibliography of al-Nadim, entitled *Kitab al-Fihrist* (*Book of the Catalog*), provides a list of all works written in Arabic by Arabic and non-Arabic authors. This bibliography is of great importance for the knowledge of mathematical works produced in Islamic countries in the 8th to 10th centuries, including Arabic translations of Greek, Syriac, and Indian texts.

the four operations, decimal fractions, sexagesimal fractions (a legacy of Babylonian mathematics), and the calculation of the square root by approximation techniques. Multiplication by 2 and division by 2, a legacy of Egyptian arithmetic, have a space of their own.

According to al-Nadim's bibliography, the second work, which we will examine more closely, is titled *Kitab al-jabr wa-al-muqabala*, that is, *Book of Algebra and Muqabala.*[15] Of the original text, there is a 1342 manuscript preserved in Oxford. The most interesting aspect for the purpose of the topic discussed in this book, namely, the use of algorithms applied to problems of everyday life, is the incipit of the work, in which the author sets out its motivation and purpose. Al-Khwarizmi says verbatim, "Caliph al-Ma'mun urged me to write a concise book on the forms of calculus with the rules of complement and reduction. I wanted the book to include what in calculus is easiest and what is most useful and what people really need in calculating inheritances, donations, division of property, judgments, commercial transactions, managing portions of land, digging canals, measurements and everything else that can be done with calculus."

Before illustrating the content and importance of al-Khwarizmi's work, it is useful to immediately clarify the meaning of the terms *al-jabr* and *al-muqabala*: the former means "complement", the latter "reduction".[16] Both terms have a definite role in the context of solving equations. The complement rule makes it possible to eliminate negative terms from equations by adding the term itself to both members. Using this rule, the equation $x^2 - 2x + 7 = 2x^2 + 3x + 3$ is transformed into the equation $x^2 + 7 = 2x^2 + 5x + 3$. The reduction rule, on the other hand, corresponds to the simplification of similar terms in the two members of the equation. According to this, the equation $x^2 + 7 = 2x^2 + 5x + 3$ is reduced to $4 = x^2 + 5x$. It is interesting to note that, as is well known, the first of the two terms,

---

[15] The meaning of the terms *algebra* and *muqabala* will be specified later. Sometimes, following Frederic Rosen's English translation dating from 1831, the title *Al-kitab al-muhtasar fi hisab al-jabr wa-l-muqabala*, i.e., Compendium of Algebra and Muqabala, is used; see Rashed, *op cit.*, p. 9.

[16] Roshdi Rashed, *L'algèbre*, in Id. (ed.), *Histoire des sciences arabes, Mathématique et physique*, vol. II, Seuil, Paris 1997, p. 31. Other authors translate the two terms as 'restoration' and 'simplification'.

algebra, has gained enormous importance, ending up by denoting the entire mathematical discipline concerned with the study and solution of equations. Later, in the last two centuries, the term algebra has assumed a wider meaning, referring to the entire field of algebraic calculus based not only on arithmetic operations but also on operators acting on abstract mathematical structures.

The work of al-Khwarizmi consists of two volumes. The first deals with the theory of first- and second-degree equations, algebraic calculus, and the solution of various problems by algebraic and geometric methods. Al-Khwarizmi calls "root" (*jidhr*) or "thing" (*shay*) the unknown, "square" or "value" (*mal*) the square of the unknown, and "number" (*adad*) the known term. At the beginning of the volume, he classifies equations into six groups that, according to the modern notation, we can represent as follows:

1. $ax^2 = bx$ 2. $ax^2 = c$
3. $ax = c$ 4. $ax^2 + bx = c$
5. $ax^2 + c = bx$ 6. $bx + c = ax^2$

For each group, he systematically presents solution methods with an algebraic formulation. Although, as noted earlier, the treatment was inspired by applications, the discussion on the resolution of the equations is carried out in an abstract and rigorous manner even though it still refers to specific numerical values (for example, with reference to a number of coins such as *dirhams*). Let us see how al-Khwarizmi treats equations of type 5[17]:

> Square plus number equals root occurs when you say: a square and twenty-one dirhams equal ten roots i.e. if you add twenty-one dirhams to a square it will equal ten roots.

This corresponds to the equation $x^2 + 21 = 10x$, for the resolution of which the author gives the following procedure:

> Halve the number of roots, the result will be five; multiply it by itself, the result will be twenty-five; subtract from it twenty-one, which we said

---

[17] Rashed, *op cit.*, p. 104.

is added to the square, that leaves four; take its [square] root which is two; subtract two from half the number of roots which is five. That leaves three; that is the root of the square you are looking for, and the square is nine. And if you will add the [square] root of four to half the number of roots, the result will be seven, which is the root of the square you are looking for and the square is forty-nine.

The solutions of the equation are therefore 3 and 7. In this case, both solutions of the second-degree equation are calculated, as they should be. It should be mentioned, on the other hand, that when one of the solutions is negative, it is ignored. Similarly, it is not considered that in some special cases the solution of an equation may be zero.[18]

On the basis of this example, we can make two considerations. The first observation is that for the formulation of the problem and the solution method, al-Khwarizmi uses what is called "rhetorical algebra".[19] As can be seen, constant values are presented in textual form (not by means of numbers), and for the operations (sums, products, calculation of square roots, etc.) no symbolic form is used such as those to which we are accustomed today. In spite of this character, the second observation we can make is that in this case, as with all the other problems presented in the text, al-Khwarizmi provides a real algorithm for calculating the solutions of the given equation. As we have said, while making use of particular numeric cases in the presentation, the author describes the resolution procedure in an abstract way, so that the procedure itself can be applied to solve other equations of the same type. In this case, expressing ourselves in modern terms, we could say that given an equation of the type $x^2 + c = bx$, the algorithm provided by al-Khwarizmi corresponds to the formula $x = b/2 +/- \sqrt{[(b/2)^2 - c]}$.

This method used by al-Khwarizmi to present the procedural steps needed to compute the solution of the equation makes us understand why his name was used in the centuries that followed to define any detailed sequence of operations needed to solve a given problem, thus giving rise to the term "algorithm".

---

[18] Of course, based on the knowledge of the time, the possibility that a solution does not exist in the real field, such as in the case of the equation $x^2 = -1$, is not even considered.

[19] Al-Khalili, *op cit.*, p. 157.

For some types of equations, al-Khwarizmi presents both arithmetic and geometric methods of solution. An example is the following (with reference to an equation of type 4)[20]:

> The square plus ten roots equals thirty-nine dirhams.

For this problem, which corresponds to the equation $x^2 + 10x = 39$, al-Khwarizmi provides two different geometric constructions. We report the second, which has been frequently cited in later texts:

> Let there be a surface *AB*, which is the square, and we want to add ten roots to it. Let us divide ten into two halves; the result will be five and we will make two surfaces on the sides of the square. Let these surfaces be *C* and *N*; the length of one of these surfaces will be five cubits which is half of the ten roots and their width is equal to one side of the surface *AB*. What remains is a square from one of the corners of the surface *AB* which is five by five. We now know that the first surface [*AB*] is the square and the two surfaces on the sides of it are the ten roots; all this makes thirty-nine. To complete the largest square there remains [to be added] the square five by five — that is, twenty-five — which we add to thirty-nine to complete the largest surface. The sum of this makes sixty-four; we take its root, which is eight, and this is the side of the largest square. If we now subtract the same amount that we had added, that is five, there remains three which is the side of the surface *AB* — that is the smaller square — three is its root and the square is nine (Figure 2).

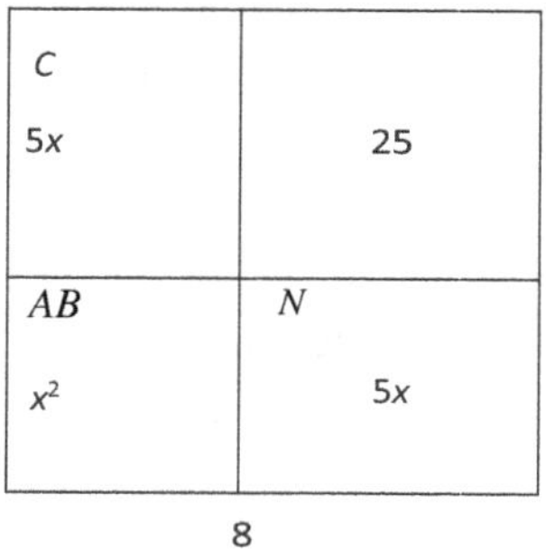

C $5x$	25
AB x^2	*N* $5x$

8

Figure 2.   Geometric construction for solving the equation $x^2 + 10x = 39$.

---

[20] Rashed, *op cit.*. pp. 108–113.

In this case, the only solution that is determined is 3. The other solution, −13, is not taken into account.

Sometimes, the presentation of the solution method ends with the sentence "And this is what I wanted to prove." The chapter devoted to equations ends as follows: "We have seen that everything you do with the calculus of *al-jabr* and *al-muqabala* necessarily leads to one of the six procedures I described at the beginning of this book of mine. I have completed their presentation; know it."

After the introductory part with a discussion on equations, the book contains several chapters dealing with multiplication, addition, subtraction, and division. Six problems are then presented that are connected to the six canonical equations; the following is the fifth problem:

Divide ten into two parts, multiply each part by itself, and sum the two products. The result is 58 dirhams.

It requires one to solve the equation $x^2 + (10 − x)^2 = 58$, which leads back to $x^2 + 21 = 10x$, i.e., the equation of type 5 with solutions 7 and 3 that we obtained previously. These examples are then followed by a chapter with 34 problems of a similar nature, which always lead back to the solution of one of the canonical equation types.

Concluding the first volume, we find two chapters of an applied nature. One is devoted to solving simple problems related to transactions (sales, purchases, exchanges, payment of wages, etc.) such as the following:

A worker is hired for ten dirhams per month; he works six days. How much compensation should he get?

The last chapter is devoted to land measurement and evidently has geometric content. In it, properties of geometric figures (e.g., the Pythagorean theorem) and rules for calculating areas and volumes are presented (with demonstrations). For π, the value $3 + 1/7 = 3.1428$ is used.

The second volume of the *Book of Algebra and Muqabala* is entitled the *Book of Wills* and contains problems of inheritance and bequests dealt with according to the rules of Islamic law. It should be noted that this issue had considerable prominence in that period since the relatively recent advent of the Islamic religion had profoundly changed the criteria previously adopted under Byzantine and Persian legislation and/or traditions. The calculation of the inheritance shares provided for male and female children and other relatives (a deceased person's husband or wife or his or her parents) is indeed quite complicated, especially in the presence of past debts or bequests to slaves or more generally to outsiders. The *Book of Wills* consists of thirteen chapters containing a long series of problems concerning bequests, granting and restitution of dowries, release of slaves, etc., with their solutions. Despite the fact that the criteria for dividing inheritances are sometimes extremely complex, the mathematics used is quite simple, since generally it consists of arithmetic operations on fractional numbers. Let us look at a particularly elementary example:

> Problem. Someone says: a woman dies and leaves a husband, a son and three daughters. She also leaves a man one-eighth plus one-seventh of her possessions. Determine the shares of the statutory division, referring to a total of twenty.

Let C be the total value of the woman's possessions. The outsider receives the value of 15/56 of C. The relatives will have to subdivide the remaining 41/56 of C. Specifically, according to the law, the husband will have 1/4 (i.e., 5/20), the son will have 6/20, and each of the three daughters will have 3/20 of the aforementioned 41/56 of C. Each share (1/20) will thus correspond to 41/1120 of the woman's total C assets.[21]

At this point, the question that arises is whether al-Khwarizmi's work is original or whether the treatment contained in the *Book of Algebra and Muqabala* derives from earlier authors. Based on an analysis of the terms used in the text, it would seem that the algebraic part is more original, while for the geometric part, the author shows familiarity with the Arabic translation of Euclid's *Elements* by al-Hajjaj, his colleague at the House of Wisdom. Moreover, it is very likely that al-Khwarizmi was also familiar with the works of the Indians Aryabhata and Brahmagupta, and he was certainly familiar, as we saw earlier, with Indian arithmetic calculus and

---

[21] *Ibid.*, p. 238.

the Sindhind astronomical system.[22] On the other hand, it seems that al-Khwarizmi did not know the works of Diophantus, which, as far as is known, were translated into Arabic a few decades after the *Book of Algebra and Muqabala* was written.

In any case, as is the case with few other mathematical works, al-Khwarizmi's text had a very important role not only in the Arab world but also in European countries for several centuries. Various translations of it exist, beginning with that by Robert of Chester made in Segovia in 1145, entitled *Liber algebrae et almucabala*, and that by Gherardo da Cremona made in Toledo, also in the mid-12th century, entitled *Liber Maumeti filii Moysi alchoarismi de algebra et almuchabala*. Subsequently, various other Latin texts by European authors took up the contents of al-Khwarizmi's work and were passed down for centuries, forming the basis of medieval mathematical thought. In particular, as noted earlier, the author's name, latinized as Algorismus, was adopted to denote, par excellence, any step-by-step computational procedure.

The Successors

During the following 10th and 11th centuries, al-Khwarizmi's work was continued in the Islamic world by a number of Arab and Persian mathematicians.

In around the middle of the 10th century, the Syrian mathematician Abu'l Hasan Ahmad ibn Ibrahim al-Uqlidisi (920–980), who worked in Damascus and Baghdad, completed the treatment of the decimal positional system. His work *Book of Chapters on Indian Arithmetics* is the oldest Arabic text that has come down to us in which algorithms for arithmetic operations are presented and in which the theory of decimal fractions is developed. In this work, a symbol corresponding to our comma is introduced to separate the integer part from the decimal part of a number. The name al-Uqlidisi seems to indicate his familiarity with Euclid's works as a translator or as an expert connoisseur of the Greek mathematician. Other authors who can be considered continuators of

---

[22] On al-Khwarizmi's knowledge of Greek and Indian mathematical literature, see *Ibid.*, pp. 30–82.

studies on the so-called "Indian calculus" are Kushyar ibn Labban, who worked in the second half of the 10th century, and Abd al-Qahir al-Baghdadi, who was active in the 11th century.

As for algebra, the new branch of mathematics identified by al-Khwarizmi, numerous mathematicians in the Islamic world devoted themselves to it after his death. Among the most influential scholars working in the second half of the 9th century, we might mention Abu Abd Allah Muhammad ibn Isa al-Mahani, who was born in Persia in Mahan, a small town in the province of Kerman, and was active in Baghdad, and who commented on the works of Euclid and Archimedes. Also of note is Thabit ibn Qurra al-Harrani, a Syriac mathematician and astronomer, a native of Harran, who drafted a new translation of Euclid's *Elements* and translated works by Archimedes, Apollonius, Nicomachus of Jerash, and Diophantus.,[23] Another scholar is Abu Kamil Shuja ibn Aslam (c. 850 — 930), an Egyptian mathematician nicknamed "the Egyptian Calculator". The latter plays a particularly important role in the history of mathematics because although he too used the style of "rhetorical algebra", he was the first to study in-depth equations of a degree greater than two, systems of multiple equations in multiple unknowns, and indeterminate equations.

Abu Kamil's works cover a wide range of computational topics. In his book also called *Book of Algebra and Muqabala*, in which he develops and systematizes the work of al-Khwarizmi, he addresses equations with fractional or irrational solutions for the first time. See, in this regard, the following example offered in the book:

> If you were asked: in an equilateral triangle the sum of its area and height is 10, how much is the height?

Assuming that the side of the triangle is x and the height is h, the problem leads to the solution of an equation of type 4: $h^2 + \sqrt{3}\,h = \sqrt{300}$.

---

[23] Al-Khalili, *op cit.*, p. 158. Ibn Qurra also wrote *Pamphlet on amiable numbers*, that is, those pairs of numbers in which each number is the sum of the divisors of the other. According to Copernicus, ibn Qurra calculated the length of the solar year with only three seconds of error.

As can be easily calculated, the solution is $h = -\sqrt{3}/2 +/- \sqrt{(3/4 + \sqrt{300})}$ and is therefore irrational.

In another work, the *Book of Rarities in the Art of Calculus*, Abu Kamil shows how to search for all solutions of an indeterminate equation. The problem of solving indeterminate equations with integer solutions was known from antiquity and was particularly popular in the Middle Ages, not only in Islamic countries but also in China and India. One work by Abu Kamil that arouses particular curiosity and always concerns indeterminate equations with integer solutions is the *Book of Birds*. The name comes from the fact that the problems dealt with in it are about buying birds of different species and at different prices with a given amount of money at hand. An example is as follows[24]: A man has to buy 100 birds with 100 *dirhams* at his disposal. Specifically, he has to buy ducks (2 *dirhams* each), pigeons (1/2 *dirham*), doves (1/3 *dirham*), larks (1/4 *dirham*), and chickens (1 *dirham*). Thus, if we denote by $x1, x2, x3, x4, x5$ the number of ducks, pigeons, doves, larks and chickens respectively, the system of equations to be solved is as follows:

$$x1 + x2 + x3 + x4 + x5 = 100$$

$$2\,x1 + 1/2\,x2 + 1/3\,x3 + 1/4\,x4 + x5 = 100$$

In this, the five unknowns must necessarily have an integer value. For this problem, Abu Kamil identifies 1,443 solutions. Among the examples presented by the mathematician, there is one for which he claims to have calculated 2,678 solutions.

Other works by Abu Kamil concern geometry: the *Book on the Pentagon and the Decagon* and the *Book on Land Measurement and Geometry*, a geometry text for non-mathematicians, a part of the results of which were taken up by Fibonacci in his *Practica geometriae*. Several other works that appear in al-Nadim's aforementioned bibliographical inventory have unfortunately been lost.

---

[24] Jacques Sesiano, *Islamic Mathematics*, in Helaine Selin (ed.), *Mathematics Across Cultures: The History of Non-Western Mathematics,* Springer, Berlin, 2001, p. 149. As we shall see in the following, the "problem of the 100 birds" was re-proposed by numerous medieval authors.

Among the continuators of al-Khwarizmi's work who were active in later centuries, we must first mention the Persian mathematician Abu l-Wafa Muhammad al-Buzjani (940–998). Born in Khorasan, he moved to Baghdad at the age of nineteen and made important contributions in various fields of astronomy and mathematics. In the area of algebra, he drafted a commentary on the works of al-Khwarizmi and translated the foundational Greek text of algebra, Diophantus' *Arithmetica*. He also devoted himself to the study of trigonometry and defined various identities related to the sine, cosine, and tangent functions (e.g., the formulas for addition and subtraction of sines and cosines). Finally, he also wrote two works of practical interest. The first is a geometry text addressed to land surveyors, entitled *Book on Geometric Constructions Necessary for the Artisan*. It consists of twelve chapters, which contain constructions and problems important for land surveying, architecture, and geodesy — calculation of surfaces, construction of squares of area equal to the sum of the areas of several given squares, or construction of polygons inscribed in a circle.[25] In the second work, entitled *Book of What is Necessary in the Science of Arithmetics for Scribes and Merchants*, there is a prevalence of applications. It does not contain demonstrations but only computation rules and examples relating to everyday life. In this text, Abu 'l-Wafa develops the calculation of fractions and, for the first time in Arabic mathematics, makes use of negative numbers.[26]

During the 11th and 12th centuries, there were two other authors who now deserve special attention: the physicist and mathematician Ibn al-Haytham (965–1040) and the mathematician and astronomer Omar Khayyam (1048–1131). Al-Haytham, who was known in the West by the Latinized name Alhazen, became famous for his treatise *Treasure of Optics*, a subject that he addressed not only from a physical and a geometrical point of view, drawing on what Ptolemy wrote about reflection and refraction, but also from a physiological point of view, that is,

---

[25]Youschkevitch, *op cit.*, pp. 108–110.

[26]*Ibid*, p. 27. It should be noted that unlike in Arabic texts, negative numbers already appeared in some Indian mathematical works (usually written in red, as opposed to positive numbers, which were instead written in black).

providing a description of the structure of the eye. Besides optics, al-Haytham devoted himself to a variety of other disciplines: mechanics, astronomy, mathematics, engineering, philosophy, and theology; he wrote, according to medieval biographers, more than two hundred works. In the mathematical field, he devoted himself to algebra, geometry, and number theory. In particular, in his work *Analysis and Synthesis*, he studied the properties of perfect numbers (i.e., numbers that are equal to the sum of their divisors), a topic already studied in Greek times by Pythagoras and Euclid. He was also the first to state that all even perfect numbers are given by the formula $2^{n-1}(2^n - 1)$, where $(2^n - 1)$ is a prime number (although proof of this fact was given by Euler many centuries later). He also defined an algorithm based on a geometric approach for calculating the sum of the first one hundred numbers.

Omar Khayyam also devoted himself to various disciplines, both scientific and literary. Born in Persia, in Nishapur, Khorasan, and of Tajik ethnicity, he is best known to most people as a philosopher and poet, author of the *Quatrains* (*Rubayyat*), although this activity was probably not his main one, but was carried out in parallel with the demanding work he did at the Isfahan Observatory or the Merv Study Center in which the Seljuk sultans, who had conquered Persia in those years, got him involved. His most important mathematical contribution concerns the study of conic sections and cubic equations. In his major mathematical work, *Treatise on the Demonstration of Problems of Algebra*, Omar Khayyam analyzes the properties of curves that are defined as conic sections (circles, ellipses, parabolas, and hyperbolas) and then turns to the study of solving cubic equations. He classifies thirteen different types of cubic equations, providing solution algorithms for each based on a geometric approach. Through appropriate transformations, he connects the solution of third-degree equations to intersections between conic sections.

Among the mathematical results Omar Khayyam claimed to have obtained, and which he had reported in work that has since been lost, is the determination of the powers of the binomial up to the sixth degree, probably with the use of the construction that we call "Tartaglia's (or Pascal's) triangle", which by that time was already known in China and knowledge of which had spread, presumably, to the Arab world as well.[27]

Omar Khayyam's work as an astronomer was also particularly important. With the use of very simple instruments — a sundial, a water clock, and an astrolabe — he was able to calculate the length of the solar year with exceptional precision — 365.242198 days (approximated as 365.2422) — which coincides with the value known today. This value corresponds to adding 8 days every 33 years (with an error in excess of 2 days every 10,000 years) and was adopted in the so-called *Jalali calendar*, which was used in Iran until the early 20th century.[28]

[27] Boyer, *op cit.*, p. 282.
[28] Al-Khalili, *op cit.*, pp. 159–160.

Chapter 3

Algorithms and Puzzles in Medieval Monasteries and Courts

Arithmetic and Religion

As we have seen, beginning in the 7th and 8th centuries and continuing until the 12th, first India and then mainly the Islamic countries were the driving force behind an intense development of the mathematical sciences, which, within a broader fervor of scientific activities that touched all areas of human knowledge, achieved results of great depth and considerable impact on the computational needs of daily life.

The history of mathematical sciences and computational methods in Western Europe is very different, however. At an early stage, after the fall of the Western Roman Empire and basically until the 9th century, arithmetic, geometry, and computational techniques were decisively neglected. To use an apt expression by the historian of mathematics Boyer, "the mathematical sciences fell into oblivion" during this period. Scholarly activity was concentrated in the monasteries and, in the wake of the mathematical works of late Roman times and the Neoplatonic philosophical schools, was reduced to a sophisticated but sterile exercise involving the analysis and classification of integer and fractional numbers, proportions, and different types of series of numbers and their properties, analyses that were intertwined with mystical and religious aspects. In this context, as we shall see, computational techniques focused on the determination of the date of Easter and a few other applications that find their rationale from being in the rural world surrounding the monasteries. Only in a later period (from the 10th and 11th centuries onward), as a result of the cultural revival taking place in cathedral schools and contact with Arabic culture, was interest in mathematics and calculus rekindled. Later, in the 12th and 13th centuries, these studies would become essential tools in the revival of mercantile activity and city administration.

It seems to us to be particularly significant that in the first phase we indicated previously, among the figures who made the most interesting contributions in the late Latin period, close to the fall of the Western Roman Empire, we do not encounter mathematicians but intellectuals and philosophers. These scholars developed their thought in a broad dimension that embraced what was generically called the "liberal arts" and which, precisely in that period, was classified as the arts of the *Trivium* (grammar, rhetoric, and dialectic) and the *Quadrivium* (geometry, arithmetic, astronomy, and music).

In the roster of these thinkers, the first we will mention is a pagan Carthaginian, Martianus Capella, who was born in Madaura and lived between the 4th and 5th centuries. In his encyclopedic work *De nuptiis Philologiae et Mercurii*, consisting of nine chapters, Capella defines and illustrates, in both prose and poetic forms, and always with an allegorical slant, the arts of the *Trivium* and the *Quadrivium*, excluding hence architecture and medicine despite the fact that they played an important role in classical Roman culture. To these disciplines it is not offered to speak at the imaginary wedding ceremony. The last four chapters of the work are devoted to the arts of the *Quadrivium*, but only in the chapter on arithmetic (Chapter VII) is there content of some interest from a mathematical point of view. In fact, Chapter VI, which is devoted to geometry, is actually predominantly a concise treatise on geography with few notions of plane and solid geometry, while the chapters devoted to astronomy and music (Chapters VIII and IX) only contain modest references to numerical aspects. The treatment of arithmetic is mainly devoted to the study of numbers and their properties and the identification of particular classes of integers (even and odd, prime and compound, mutual primes, perfect numbers, etc.) without any reference to algorithmic or computational aspects.

Another person who left an important mark on early medieval mathematics was the scholar and philosopher Severinus Boethius (480–524). Although his preeminent activity was as a Roman senator and adviser at the court of the Ostrogothic King Theodoric (until he came into conflict with Theodoric and was therefore imprisoned, tried, and executed), Boethius had a very intense cultural life and wrote a large number of literary and philosophical works, the most famous of which is *De consolatione philosophiae*, which he wrote in prison. Within his extensive literary output, which includes works on philosophy, logic, and theology, Boethius also devoted himself to the arts of the Quadrivium, among which he established a precise hierarchy that saw arithmetic in first place, geometry in second place, then music, and finally astronomy. This order was established on the basis that the latter disciplines need the former ones, while the former disciplines can also exist independently For example, geometry needs arithmetic, but the reverse is not true. So too, to be understood and practiced, music needs arithmetic (particularly number theory), but arithmetic does not need music. This hierarchy derives from the Neoplatonic conception that "there are degrees to be traveled in the ascent to truth, but it is necessary to begin with that which constitutes its principle, namely arithmetic."[1]

Much of what we know about Boethius we owe to his most important pupil, Cassiodorus (c. 490–580). Flavius Cassiodorus was a Roman politician, scholar, and historian who worked first under the Roman-Barbarian reign of the Ostrogoths and later in the Eastern Roman Empire. He reported that during a trip to Greece, Boethius devoted himself to studying and translating *Arithmetica* written by Nicomachus of Gerasa[2] (a Greek mathematician who lived from about the year 60 to 120), as well as Ptolemy's *Almagest* and the first book of Euclid's *Elements*. Within the arts of the Quadrivium, Boethius' works certainly

---

[1]Ambrosetti, *L'eredità Arabo-Islamica*, p. 13.

[2]Nicomachus of Gerasa belonged to the Greek Pythagorean school. In his work *Arithmetica*, he devoted himself to the study of the properties of numbers as well as their philosophical significance. It was he who argued that arithmetic is the basis of geometry, music, and astronomy, a concept that influenced the definition of the arts in the Quadrivium. Nicomachus' thought and his numerological studies deeply permeated early medieval culture through the work of Boethius.

include *De institutione arithmetica* and *De institutione musica*; however, based on what Cassiodorus reports, we must assume that Boethius also wrote texts devoted to geometry and astronomy. His treatise *De institutione arithmetica* is heavily influenced by the work of Nicomachus. In it, Boethius develops a thorough classification of numbers and their properties, beginning with even and odd numbers; continuing with prime numbers (to identify which he illustrates Eratosthenes' sieve), compound numbers, and perfect numbers; and introducing several other classes of integers endowed with special properties. He also devotes himself to the study of proportions and series, drawing on Pythagorean concepts (triangular, square, and pentagonal numbers) and emphasizing the interpretation of ratios between integers as a musical function. However, as in the case of Martianus Capella, the whole work does not have algorithmic aspects; rather, the references to Pythagoras tend to associate mystic and religious aspects with numbers. For example, on the basis of the properties of numbers, it is argued that the creation of the world took six days because 6 is a perfect number (that is, equal to the sum of proper divisors) or that 7 is the number of the Virgin because, among the first ten numbers, it is the only one that does not divide any and is not divided by any other.

Cassiodorus also considers arithmetic the foundation of the arts of the Quadrivium. In his work *Institutiones divinarum et saecularium litterarum*, he succinctly sets out a classification of numbers according to their properties, following the definitions introduced by Martianus Capella and Boethius. Like his predecessors, Cassiodorus emphasizes the presence of numbers in the Bible, thus helping to attribute a mystical significance to numbers that characterized Medieval thought and art. More generally, according to this approach, it is all of culture and all of education that are measured by the need to understand and analyze the truths revealed in Holy Scriptures.[3] Compared to Boethius, however, Cassiodorus emphasizes that the importance of numbers lies not only in their numerological meaning and abstract properties but also, above all, in their many applications in everyday life and in the possibility of

---

[3] Ambrosetti, *op cit.*, p. 25.

developing calculations (for example, in commercial and financial accounting and in the measurement of time).[4]

In fact, it is precisely the measurement of time for religious and liturgical purposes that became the main reason for the study and development of computation in these centuries. At the same time, philosophical, literary, and scientific schools elected monasteries as their headquarters, and *computus* — that is, the calculation of time aimed at determining the date of Easter — became one of the topics on which mathematical studies focused. As we have already noted, the construction of calendars and the computation of the date of Easter (in this case, Passover) played an important role in mathematics even in al-Khwarizmi's time and maintained this importance throughout the medieval centuries, constituting one of the keys linking arithmetic and religion.[5]

The determination of the date of Easter was also complex because it had to meet several conditions. The first two are related to the remembrance of Passover: that is, the day when the people of Israel left Egypt to return to the Promised Land under the leadership of Moses. According to the Bible, this event happened on the day of the first full moon after the spring equinox. In addition, another constraint for the date of Easter comes from the fact that it must fall on a Sunday, the day of Christ's Resurrection. The need to combine the different constraints first required that the date of the spring equinox (initially believed to be March 25) be defined once and for all. In 325, the Council of Nicaea established the date still in use today, 21 March, as the date of the spring equinox. Secondly, it was necessary to combine the length of the lunar month (29.5306 days) and the aforementioned nineteen-year Metonic cycle with the length of the solar year (365.2422 days). All of this, without particularly sophisticated mathematical skills, required some ability in arithmetic to the point that the calculation of the date of Easter (*computus Paschalis*) became a the calculation (*computus*) par excellence.

---

[4]Teun Koetsier and Luc Bergmans (eds.), *Mathematics and the Divine. A Historical Study*, Elsevier, Amsterdam, Philadelphia, 2005.

[5]Another meeting ground between religion and arithmetic was music, as elaborated in detail by Boethius in his work *De musica*; see Koetsier and Bergmans (eds.), *op cit.*, pp. 18–19.

A particular echo of the study of *computus* is found in the monasteries of the British Isles. According to some historians, the evolution of monasteries in an intellectual sense has its origin precisely in Ireland and the Anglo-Saxon kingdoms of Great Britain perhaps because of the absence of antagonism to a particularly developed urban intellectual reality.[6] As we shall see, this spirit will be reflected a century later in the intellectual role monasteries would assume in the Carolingian kingdom. In 658, an anonymous author, an Irish cleric, compiled a work entitled *De computo dialogus*. This text, which deals with the subject of arithmetic and the computation of time precisely in the form of a dialogue between a *magister* and a *discipulus*, opens with a quotation from St. Augustine about the importance of numbers as they constitute one of the four foundations of Sacred Scripture along with divine law, history, and grammar. This affirmation of the conceptual primacy of numbers, and thus of arithmetic, is then reinforced by a second quotation, from Isidore of Seville[7] which still appears to be a eulogy of mathematical rationality: "Indeed, thanks to numbers we do not fall into confusion. If you take away numbers everything falls into ruin. If you take calculation out of the world blind ignorance will envelop all things; he who ignores calculation cannot be distinguished from other animals."[8] According to the author, numbers were discovered by Abraham and then arithmetic was transmitted to the Egyptians by Moses. He states that among the Greeks, the main devotee of arithmetic was Pythagoras, and among the Latins, it was Boethius. In the course of its treatment, *De computo dialogus* enriches the subject of calendars with developments that tread a fine line between religion, philosophy, astronomy, and mathematics, although the real mathematical content is at a very elementary level.

---

[6] Albrecht Diem, The Emergence of Monastic Schools. The Role of Alcuin, *Proceedings of the Third Germania Latina Conference*, Groningen 1995, p. 33.

[7] Isidore of Seville (Isidorus Hispalensis, c. 560 636), who is now venerated as a saint and doctor of the Church, was bishop of the Spanish city during the reign of the Visigoths and was the author not only of many literary (such as a collection of synonyms) and historical works (*Historia de regibus Gothorum, Wandalorum et Suevorum*) but also a work of an encyclopedic nature (*Etymologiarum sive Originum libri XX*) in which he dealt with the topics of arithmetic, geometry, music, and astronomy.

[8] Koetsier and Bergmans (eds.), *op cit.*, pp. 186–187.

The work of the English monk Bede, known as the Venerable (672–735), developed along deeper and more challenging lines. Bede was led at the age of seven to the Benedictine monastery of St. Peter's in Wearmouth, Northumbria, in whose territory he was born, as he himself relates. He spent his entire life first in this monastery and later, until his death, in a sister monastery in Jarrow. In addition to performing monastic duties and devoting himself to study and teaching, Bede was the author of an extensive output in a variety of fields. The web portal *Documenta Catholica Omnia* lists about seventy of his works (as well as dozens of texts of uncertain attribution). In particular, he drafted treatises of historical content (*Historia ecclesiastica gentis Anglorum*, in which there are also brief autobiographical notes), works of a theological nature (his editions of the Gospels and the Bible enjoyed great fame and were used until the 19th century), as well as literary (works on grammar, spelling, etc.) and scientific works (*De rerum natura*).

With regard to the measurement of time, the construction of calendars, and the drafting of chronologies, several of Bede's writings can be identified that cover these themes, in particular *Computus vulgaris qui dicitur ephemeris*, in which data on the phases of the moon are provided, and *De ratione computi*, in which the question of the measurement of time (hours, days, weeks, etc.) and the subject of the nineteen-year cycle are addressed in didactic dialogic form. The short treatise *De numerorum divisione libellus* is devoted to arithmetical operations and the measurement of distances. In it, the units of measurement of distance used by the Romans (finger, ounce, foot, step, pole, stadium, league, etc.) are studied and the relations among them illustrated.[9] Finally, in *De temporibus*, after

---

[9] In this work, Bede also reports a fairly realistic assessment of the circumference of the Earth, which he estimates at 252,000 stadia (a value already calculated by Eratosthenes; see Boyer, *Storia della Matematica*, p. 199). Since in the system of measurement inherited from the Romans used by Bede the foot corresponded to 29.6 centimeters, and in *De divisione numerorum libellus* he states that the stadium corresponded to 625 feet, we can deduce that one stadium corresponded to 185 meters, and therefore the circumference of the Earth reported by Bede is 46,620 kilometers. In fact, it seems that Eratosthenes was actually referring to the Egyptian stadium, which measured between 155 and 160 meters, and thus the measurement given by Eratosthenes corresponds to about 39,500 kilometers, much closer to the actual value.

a brief discussion on the measurement of time similar to that contained in *De ratione computi*, a concise history of the world from Adam to Bede's own time is presented.

Bede's most important work in this area, which he completed in 725, is *De temporum ratione*, in which the same themes as those in *De temporibus*, which was written some twenty years earlier, are taken up and developed in depth in seventy-one chapters. The text has an appendix containing a table of the dates of Easter from 532 to 1063. The first four chapters are a preparation for *computus* and concern the digital calculus[10] (Figure 1), the measurement of time, and the measurement of space. Subsequent chapters (5 to 40) deal extensively with the subdivision of time into days, weeks, and months, developing in particular the denomination of months in different cultures, the constellations, the zodiac, the phases of the moon, the tides, equinoxes and solstices, shadow length at different stages of the year and at different latitudes, the four seasons and seasonal diseases, and the five climate zones of the Earth (Arctic, temperate, torrid, austral temperate, and Antarctic). Chapters 41 to 62 deal with the study of the phases of the moon, leap years, and the nineteen-year cycle. Finally, from Chapter 64 until the conclusion, the work presents an extended history of the six ages of the world from Adam to the 8th century, developing what had already been set forth in *De temporibus* and concluding with a projected look at the following ages until the end of the world, the coming of the Antichrist, and universal judgment.

As mentioned earlier, the computational aspects of Bede's work are somewhat limited. They are concentrated in the chapters on digital calculus, in which he also provides an interesting method of using calculus itself as a technique for secretly encoding verbal texts (taking reverse advantage of the encoding of numbers by letters of the Greek alphabet), and in the chapters on the calculation of lunar cycles and the determination of the date of Easter. However, as a whole, the work of Bede (who

---

[10] In the first chapter, entitled *De computo vel loquela digitorum*, Bede takes up the method of representing the numbers from 1 to 9,999 according to the position of the fingers of the hands, a system that, as noted earlier, was already in use among the Romans and various peoples of the Middle East.

Figure 1. The digital calculus in the depiction given by Raban Maurus in his work *De Numeris* (9th century CE). (Instituto da Biblioteca Nacional, Lisbon, Portugal/ Bridgeman Images.)

was proclaimed a saint and doctor of the Catholic Church in 1899) remains a pillar of English medieval literature and a significant testament to the astronomical and geographical knowledge and calculation skills of the 8th century. Moreover, Bede's legacy was picked up by an important figure who played a notable role in the so-called Carolingian revival: the philosopher Alcuin.

Alcuin and the Mathematical Education of the Young

Among the Venerable Bede's pupils, the most influential was Egbert of York. Egbert was appointed Bishop of York in 732 and Archbishop in 735. Two years later, his brother Eadberht became king of Northumbria. Thanks to his collaboration with him, Egbert established a school at the See of York aimed not only at young clergymen but also at the sons of the nobility. Among other things, the school in York was equipped with an important library, which quickly acquired great notoriety; according to some historians, it was unparalleled in the Western world of its time. After Egbert, two of his pupils had the task of leading the school: Albert and, from 767, Alcuin.[11]

Alcuin of York (c. 732–804), who is also known by his Anglo-Saxon name Alhwin or Alchoin or his Latin name Albinus Flaccus, was born in Northumbria, probably to a family of minor nobles. When he attended school in York, he was taken under the tutelage of Egbert, and after completing his studies became a teacher at the same school. As mentioned, he was appointed principal of the school in 767 and became a deacon at around the same time. He was never ordained as a priest, but nevertheless lived a monastic life, particularly when, from 796 to 804, the date of his death, he became Abbot of the monastery of St. Martin in

---

[11] Both Albert and Alcuin enriched the holdings of York's library, which initially contained only books of liturgical interest, with works collected on the European continent. In a poem, Alcuin himself recalls Egbert's travels in search of books as he had accompanied him on this sort of pilgrimage at least once; see Mary Garrison, *The Library of Alcuin's York*, in Richard Gameson (ed.), *Cambridge History of the Book in Britain*, vol. I, Cambridge University Press, Cambridge, 2019, pp. 633–665.

Tours. In 781, he had the occasion to meet Charlemagne during a trip to Italy. The Frankish king was impressed by Alcuin's learning and the expertise he had acquired in leading the school in York and suggested that he travel to Aachen, one of the capitals of Charlemagne's itinerant court, to lead the *Schola Palatina*. The prestigious institution had been created by the king's predecessors for the education of the youth of the royal family, but Charlemagne intended to transform it into a true cultural institution in which the liberal arts of the Trivium and Quadrivium were studied according to the teachings of Cassiodorus. Indeed, it is important to keep in mind that in those years, Charlemagne had initiated a process of economic and cultural revival of his kingdom. The cornerstones of this policy included on the one hand the regulations contained in *Capitulare de villis*, which created interesting spaces for the agricultural and commercial development of the countryside, and on the other hand intense international relations, primarily those with the caliph Harun ar-Rashid, which opened Christian Europe to the influences of Arab science.

Under Alcuin's leadership, the school became a reference point for many intellectuals, who flocked there from various territories of the kingdom, and collaborators, whom Alcuin called from England. Charlemagne's sons Pepin and Ludwig and the king himself attended the school, which became known as the School of Master Albinus and was open to young clergymen and the sons of noble families. Alcuin was also credited with promoting the establishment of schools at cathedral churches, thus helping to bring culture out of the narrow confines of monasteries and setting the stage for a more open educational system. It should be considered that in 815, the School of St. Germain in Auxerre housed six hundred monks and nearly five thousand students. An important document that shows the breadth of vision of the cultural initiative desired by Charlemagne and led by Alcuin is *Epistola de litteris colendis* (which was probably written by Alcuin himself), which was sent by Charlemagne to personalities representing culture within his kingdom. In it, we read the following: "Although proper conduct is more important than knowledge, nevertheless knowledge precedes conduct." And again, "Everyone should study the disciplines to which he wishes to apply himself, so that the mind may know fully what is to be done and the tongue can devote itself to praising Almighty God without the risk of error."[12]

A few years later, in around 790, Charlemagne promulgated the capitular *Admonitio generalis*, another fundamental piece of Carolingian policy. Chapter 72, which is part of the section on the clergy, and was also probably drafted by Alcuin, contains norms that were to inspire ecclesiastical and monastic life from an intellectual point of view. This text, too, makes it clear that among the priorities of Carolingian politics was the raising of the cultural level, especially of those who would later have the responsibility of educating the young. It is in this context that religious figures were urged to found *scholae legentium puerorum*, and a list of subjects that were to be taught was also provided, including *computus*.[13]

Alcuin had vast cultural knowledge. Einard describes him as "the most learned one can meet anywhere" and "the most learned in every field." The web portal *Documenta Catholica Omnia* lists about forty of Alcuin's works (some of uncertain attribution), ranging from theology to philosophy and from literature to history. In particular, he wrote many works of a theological-philosophical nature — including *De animae ratione, De virtutibus et vitiis ad Widonem comitem*, and *De fide sanctae et individuae Trinitatis* — and of a literary nature — such as *Ars grammatica, De retorica, De ortografia*, and *De dialectica*. On the topic of the history of religion, he wrote about the lives of various saints and blessed people, as well as a history of the Church of York. He also devoted himself to poetry (*Carmina*). Finally, his letters (*Epistolae*), three hundred and eleven of which refer to his correspondence with Charlemagne, are a record of enormous historical and cultural value.

One of the most unique works that made Alcuin famous and shows his versatility and dedication to the education of young people is *Propositiones ad acuendos iuvenes* (Problems to sharpen the young),

---

[12] Alfred Boretius (ed.), *Epistola de litteris colendis*, in *MGH Capitularia regum Francorum*, vol. I, Hanover, 1883, pp. 78–79; see Rurger Kramer, 'Ecce Fabula!' Problem-Solving by Numbers in the Carolingian World: The Case of the Propositiones ad Acuendos Iuvenes, *Proceedings of the MEMSA Student Conference*, 2015.

[13] Diem, *op cit.*, p. 36.

a collection of fifty-six logical-mathematical problems. Thirteen manuscripts of this text are still available (the oldest from the 9th century).[14] The question of the attribution of the text has never been fully resolved. One of the versions, published in 1563, is attributed to Bede, but there are at least two indications that this attribution is incorrect. In fact, the preface (*Monitum praevium*) of the text *Propositiones Alcuini Doctoris Caroli Magni Imperatoris ad acuendos iuvenes*[15] makes it clear that Bede himself *nullam mentionem fecerit in catalogo suorum Operum quem ipsemet confecit* (that is, there is no mention of this work in the catalogue that Bede himself wrote). Furthermore, a passage from Epistle 101, sent by Alcuin to Charlemagne in 799, is quoted as saying *misi excellentiae vestrae [...] et aliquas figuras arithmeticae subtilitatis, laetitiae causa* (I also sent a few pages of arithmetical subtleties to your Excellency for your pleasure). Other scholars speculate that the text was composed by a monk from the monastery of Chabanais named Ademar (or Aymar), who lived from 988 to 1030.[16] The prevailing view, however, is that it was actually drafted by Alcuin.

Propositiones presents a variety of simple mathematical problems, essentially derived from aspects of everyday life. Many consist of solving first-degree equations in one unknown; some concern first-degree equations in two unknowns, determinate or indeterminate; some are geometric in nature; and finally, some are logical problems. The proposed problems are accompanied by their respective solutions, but generally no detailed description of the solution method followed is provided; only rarely is the outline of an actual solution algorithm presented. Let us look at some examples[17]:

III. PROPOSITIO DE DUOBUS PROFICISCENTIBUS. Two men walking in the street see some storks and say to themselves, how many

---

[14] Franci (ed.), *Alcuino di York*, pp. 23–24.

[15] Jacques-Paul Migne (ed.), *B. Flacci Albini seu Alcuini abbatis et Caroli magni imperatoris magistri opera omnia*, 2 vols., Paris, 1851 (reprinted Hachette BnF, Paris, 2018).

[16] Peter J. Burkholder, Alcuin of York's "Propositiones ad Acuendos Iuvenes", *Propositions for Sharpening Youths*, Introduction and Commentary, in *Electronic Bulletin for the History and Philosophy of Science and Technology*, vol. 1, no. 2, 1993.

[17] Franci (ed.), *op cit.*

are they? Confabulating among themselves about the number, they said: If they were as many and three times as many and then half the third and added 2 there would be 100. Tell, whoever is able, how many there are that were seen by them at first.

This is a problem of the type called "heap problems" which, as we have already seen, can be solved by the false position method or, more simply, by a direct algebraic method. The equation that corresponds to the statement of the problem is $x + x + x + x/2 + 2 = 100$, where x denotes the number of storks seen by the two wayfarers. On solving the equation, we find $x = (98 \times 2)/7 = 28$.

XII. PROPOSITIO DE QUODAM PATERFAMILIAS ET TRIBUS FILIIS EJUS. A father died, and bequeathed to his three sons 30 glass cruets, ten of which were filled with oil, another ten half-filled, the third ten empty. Divide, whoever can, oil and cruets so that each of the three sons gets the same amount of both glass and oil.

According to historian Franci, this type of problem was very popular in Europe, and no non-European versions are known. In this case, several solutions are possible and can be obtained easily with an exhaustive search algorithm. If we denote the three siblings by A, B, and C and give a vector with the number of cruets he receives for each of them (full, half-full, empty), we find that the solution proposed by Alcuin is $A = (0, 10, 0)$, $B = (5, 0, 5)$. $C = (5, 0, 5)$. The scalar product of each of the three vectors by the vector $(1, 1, 1)$ gives us the total number of cruets each receives, and the scalar product for the vector $(1, 1/2, 0)$ gives us the amount of oil (expressed in cruets) for each. Alcuin does not indicate that there may be other solutions such as $A = (5, 0, 5)$, $B = (4, 2, 4)$, $C = (1, 8, 1)$ or $A = (4, 2, 4)$, $B = (3, 4, 3)$, $C = (3, 4, 3)$.

XXV. PROPOSITIO DE CAMPO ROTUNDO. There is a round field that has the circumference of 400 perches. Say how many *arapenni* it must hold.

This is a geometric problem in which one is asked to calculate the area of a circle (A) given the circumference (c). In addition, one must use as the unit of area the *arapenno*, corresponding to the area of 12 × 12 perches. One must then first calculate the area in square perches with the expression $A = c^2/4\pi$ and then relate that area in *arapenni*, dividing A twice by 12. Alcuin sometimes uses 3 and sometimes 4 as a value of π (thus ignoring the much more precise values determined since antiquity). In the text cited earlier, the value 4 is used and $A = 10,000$ is obtained. Dividing twice by 12 gives 69.4 *arapenni* (which Alcuin approximates to 69).

> XXXVIII. PROPOSITIO DE QUODAM EMPTORE IN ORIENTE. In the East, a man wanted to buy 100 animals of various species with 100 coins. He orders one of his servants to buy a camel for 5 coins, a donkey for 1 coin and 20 sheep for 1 coin. Say, whoever you like, how many camels, how many donkeys and how many sheeps there were in that business of 100 coins.

This problem is of the "100 birds" type. Alcuin's work presents this type of problem for the first time in the Western world. As we have seen, such problems had previously been proposed in works of Arabic mathematics. In this case, it gives rise to the following two equations (in which the three unknowns c, d, and s are the numbers of camels, donkeys, and sheeps purchased, respectively):

$$c + d + s = 100$$

$$5c + d + s/20 = 100$$

Deriving the value of $d = 100 - c - s$ from the first equation and substituting it in the second equation gives $80c = 19s$, which can be solved by assuming $s = 80$ and $c = 19$. Substituting these values into the first equation gives $d = 1$. This type of problem generally gives indeterminate equations: that is, they can admit multiple solutions, integers by assumption. In this case, the only other possible solution is $c = 0$, $d = 100$, $s = 0$, but in Alcuin's text, it is not considered. In fact, if we consider that the expression "100 animals of various species" represents the constraint that at least two of the unknowns must be different from 0, there are no other possible solutions besides the one given.

Other kinds of problems presented in *Propositiones* are of the arithmetic type, requiring simple multiplication or division, such as the problem involving adding up the first hundred numbers or calculating the powers of 2 up to the thirtieth; chase problems, involving calculating how long a faster animal will reach a slower animal; and will problems, which are analogous to those seen in the Arabic mathematical works. In addition, it is particularly interesting that Alcuin also includes in his text problems of a logical type: that is, ones that do not require arithmetical operations but simply the use of reasoning, such as the problems of crossing a river with a boat that can carry only two people, among which the problem of the three men with their sisters (renamed in later centuries as the problem of the "three jealous husbands") has had considerable success. Other problems that we might call of a logical nature are those related to strange kinship, determined, for example, by the marriages of a widower to a woman and of his son to her mother. Even more uniquely, but interesting in an educational manner, the collection also contains a problem that admits no solutions (substantially, the problem requires dividing an even number into three odd numbers).

Considering that the typology of the problems is not always original but includes mathematical problems faced over the centuries or even found in Babylonian and Egyptian mathematics, one can imagine that Alcuin drew inspiration from texts contained in the well-stocked libraries of York or Aachen or from information received from Arab scholars he met in the ambassadorships that were exchanged between Charlemagne and Caliph Harun ar-Rashid. Information may also have circulated through the intermediary of the Byzantines, with whom the Carolingian court had frequent contact. On the whole, however, it appears that about 20 percent of the problems posed are original, and, according to recreational mathematics scholar David Singmaster, such a presence of original queries is "an extraordinary proportion,"[18] in the context of collections of recreational mathematical problems.

As we have seen, the problems posed in *Propositiones* are related to everyday life. In particular, given the monastic life led by Alcuin,

---

[18] Franci (ed.), *op cit.*, p. 25. See also David Singmaster and John Hadley, Problems to Sharpen the Young, in *The Mathematical Gazette*, vol. 76, no. 475, 1992.

especially in the last years of his life — that is, in the very period in which *Propositiones* was probably written — we are not surprised that the aspects of daily life that emerge in the problems are related to the rural world of central Europe. The protagonists of the events presented are wayfarers, peasants, masons, schoolchildren, priors, and monks; the coins are denarii, solids, and talents, and are made of gold, silver, tin, and *oricalco* (a special type of brass used for coins, consisting of 90 percent copper and 10 per cent zinc); and the foods are wheat, bread, wine, oil, and eggs. The animals mentioned are horses, oxen, sheep, dogs, and wolves, and especially pigs (boars, sows, and piglets are mentioned repeatedly, suggesting that pork was particularly present among the foods); camels are also mentioned several times[19]; and the birds are doves, swallows, and storks. A picture emerges of the reality of the French countryside.

The Philosophers' Game

For a few centuries during the Middle Ages, recreational mathematics and the study of problems and games that required logical and computational skills found devotees within monasteries. An interesting example of this type of activity is the game called *rithmomachia* (from the Greek, "battle of numbers"), also known as the "philosophers' game". Its rules are quite complex and recall some of the notions found in Boethius' work. According to the historian Nadia Ambrosetti, "the spread of this game is an indirect demonstration of the continuation of Boethius' teaching of arithmetic."[20] More generally, it is said that this game aimed to educate young people to understand the harmony of the universe through arithmetic.

The origins of *rithmomachia* are not known. Although there are no traces of it in Greek literature, its name might suggest that it originated in the mathematical schools of Constantinople or Alexandria in the early centuries of our era. Some authors have attributed its invention even to

---

[19] Perhaps the presence of camels in Alcuin's text may come as a surprise, but it seems that in Carolingian times there was a custom of purchasing these animals in the East for use on long-distance journeys, including for military purposes; see Franci (ed.), *op cit.*, p. 101.
[20] Ambrosetti, *op cit.*, p. 31.

Pythagoras. The first written mention related to *rithmomachia* is traced back to a monk, named Asilo, who in 1030 conceived a game, the purpose of which was to teach his monastery's students the principles contained in Boethius' treatise *De institutione arithmetica.*[21] A few years later, a monk from the monastery of Reichenau, Hermannus Contractus (1013–1054), further developed the rules of the game. During the 11th and 12th centuries, *rithmomachia* spread rapidly in France and Germany.

A detailed description of the game would be beyond the scope of this book. We therefore provide only a few hints to show what computational skills were required of the participants.[22] The game was played between two people placed at the ends of a double chessboard (8 × 16). Each player had 24 pawns: 8 round, 8 triangular, and 7 square, and one particular pawn called a *pyramid* (or, in some descriptions of the game, a *king* because of its importance), consisting of the superposition of several round, square, and triangular pawns. The pawns were placed at either end of the board. One of the players used white pawns and the other used black pawns, but all the pawns were white on one side and black on the other because when they were captured, they were not eliminated but passed into the opposing camp. All the pawns bore a number. The four round tokens in the first row of one array (called the "even-numbered array") bore the numbers 2, 4, 6, and 8, and the round tokens in the second row bore the square of those numbers: 4, 16, 36, and 64; for the other array (called the "odd-numbered array"), the first row bore the numbers 3, 5, 7, and 9 and the second row the squares 9, 25, 49, and 81. The other tokens were numbered according to appropriate calculations inspired by Boethian rules. For example, the triangular token that was placed below the two round tokens of values 6 and 36 at the beginning of the game took the value 36 + 36(1/6) = 42 which with respect to 36 satisfied the ratio $(n^2 + n)/n^2 = (n + 1)/n$ for $n = 6$. Thus, the other triangular tokens also assumed values that respected these proportions with respect to the above round tokens, proportions that Boethius calls

[21] David Sepkoski, Ann E. Moyer: The Philosopher's Game: Rithmomachia in Medieval and Renaissance Europe, in *Isis*, vol. 95, no. 4, 2004, p. 698.
[22] David Eugene Smith and Clara C. Eaton, Rithmomachia, the Great Medieval Number Game, in *The American Mathematical Monthly*, vol. XVIII, no. 4, 1911, pp. 73–80.

superparticulares.[23] The pyramid of the even-numbered array had the value $1 + 2^2 + 3^2 + 4^2 + 5^2 + 6^2 = 91$ (since 6 is a perfect number — that is, equal to the sum of its divisors — it was called a *pyramide perfecta*) and the other had as its value the sum of the squares of the numbers 4 to 8 – that is, 190 – and was called *pyramide tricurta* because its value was without the squares of the numbers 1, 2, and 3.

The entire dynamics of the game, namely, the capture of the adversary's pawns (there were four modes of capture) and the situations that determined victory (in the best-known variant of the game, there were eight modes of victory), were developed according to arithmetic criteria. Players took turns moving a pawn. Round tokens could move on the board by one square in the four orthogonal directions, triangular tokens by two squares in the four diagonal directions, square tokens could take three steps in the four orthogonal directions, and pyramids could move like the queen in chess, but only by four squares at most.[24] The four modes of capture (by encounter, by assault, by ambush, or by siege) were determined by mathematical rules. In the simplest case, encounter, a pawn of value n could be captured by another pawn of value m greater than or equal to n if it could be placed on the same square of the chessboard by moving it according to ordinary rules. In the case of the assault, however, if k empty squares existed between a token of value n and one of value m greater than n, the token of value m could be captured if $kn = m$. In any case, the capturing token remained in the position it occupied before the move, and the captured token was turned over and acquired by the opposing team.

As can be noted, the players' ability to quickly carry out arithmetic calculations to choose the most promising game strategies was crucial. This was particularly true in order to be able to claim victory. This actually depended on the numerical properties satisfied by the values associated with the tokens of the two players. It was generally expected that

---

[23] Instead, the proportions that were used to assign value to square tokens were the proportions *superpartientes*: $A/B = (2n + 1)/(n + 1)$.

[24] Jorge Nuno Silva, Teaching and Playing 1000 Years Ago. Rithmomachia, in Costantino Sigismondi (ed.), *Orbe Novus. Astronomia e studi Gerbertiani*, Universitalia, Rome, 2010, p. 145.

surviving pawns of a particular color and captured opposing pawns would meet some particular property set at the beginning of the game. For example, if it was decided that 160 was the winning value, the game would be won by the first player who managed to capture opposing pawns the sum of whose values reached or exceeded 160. The most interesting victory options in the classic version of the game ("proper" victories: *magna*, *major*, and *praestantissima* or *excellentissima*) were based on more complex arithmetic rules related to the existence of progressions (arithmetic, geometric, and harmonic[25]) between values associated with aligned tokens. *Victoria major* was achieved with four aligned tokens that, when taken in groups of three, respected the arithmetic and geometric progressions; for example, with the four tokens (2, 3, 4, and 8), the arithmetic progression (2, 3, and 4) and the geometric progression (2, 4, and 8) are respected. In contrast, *victoria excellentissima* required that all three progressions be verified, as is the case with the four values (4, 6, 9, and 12) respecting the arithmetic (6, 9, and 12), the geometric (4, 6, and 9), and the harmonic (4, 6, and 12)[26] progressions.

The game thus required considerable calculation skills. Although the arithmetic operations required were simple enough, it should be kept in mind that, when the game was invented, it involved operating with numbers in Roman notation; it would not be until two hundred years later that Indo-Arabic positional notation would begin to take hold. For this reason, it seems that players always had a Pythagorean table at hand.

Rithmomachia maintained a certain notoriety for many centuries. In the 13th century, the mathematician and physicist Jordanus Nemorarius (1225–1260) the author of several treatises on statics and algebra, devoted a small essay to the game. During the Renaissance, Thomas More imagined the game being practiced in his hypothetical world of Utopia. In 1539, a Florentine named Benedetto Varchi sent two texts from Padua to Florence: one relating to this game, which he called the *game of Pythagoras*, and one concerning the proportions adopted in it to assign

---

[25] A series of numbers is in harmonic progression if their reciprocals are in arithmetic progression.

[26] Jeff Suzuki, *Mathematics in Historical Context*, Mathematical Association of America, 2009.

value to the pawns.[27] A very accurate and detailed description of *rithmomachia* was written in Elizabethan times, in 1563 (it was probably an English translation of a French text by Boissière from 1554). In the text, the game is presented as follows: "The noblest, most ancient and erudite game, called the Game of the Philosophers, invented for the honest recreation of students and other persons, in order to overcome the boredom of time, to rest from work and exercise cunning."[28]

[27] Franco Pratesi, Il gioco dei filosofi fiorentini, in *L'Italia Scacchistica*, no. 7, 1995, pp. 171–173.
[28] Ann E. Moyer, *The Philosopher's Game: Rithmomachia in Medieval and Renaissance Europe, with an Edition by Ralph Lever and William Fulke - The Most Noble, Ancient and Learned Playe (1563)*, The University of Michigan Press, Ann Arbor, MI, 2001.

Chapter 4

The Penetration of Indo-Arabic Numbers in the West

The Abacus of Gerbert d'Aurillac

Among the various hypotheses that have been put forward regarding the invention of the *rithmomachia* game, there is also the one (which has not been corroborated by any sure evidence) that it was created, or at least improved, by a person who has gone down in history as the first (and perhaps the last) mathematical pope and as one of the first Western scholars to have grasped the great importance of the positional system based on the nine Indo-Arabic digits, Gerbert of Aurillac, who was appointed pope in 999 under the name of Sylvester II.

Gerbert was born in Aurillac, Auvergne, around 945. His life is an impressive series of political, ecclesiastical, educational, and cultural assignments that he led with determination and intelligence until he reached the papal throne.[1] As an oblate at the age of thirteen, he entered the convent of St. Gerald of Aurillac, where he was instructed in the arts of the Trivium and Quadrivium and was ordained a Benedictine monk. Gerbert was very capable and studious, and when the Count of Barcelona, Borrell, and the bishop of Vich (a town in the province of Barcelona), Attone, visited the convent of St. Gerald, they decided to take him with them to Spain. There, in the years between 967 and 970, Gerbert had the opportunity to read the classical Greek and Latin texts kept in the Abbey

---

[1] P. Riché, *Gerbert d'Aurillac. Le pape de l'an mil*, Fayard, Paris, 2006.

of Santa Maria de Ripoll, study the philosophical and mathematical works of Boethius and Isidore of Seville, and come into contact with Arab culture, which enabled him to learn about the astrolabe and the methods of calculation based on Indo-Arabic numerals. During a trip to Rome, he met Pope John XIII and the emperor of the Holy Roman Empire (as well as the king of Italy), Otto I of Saxony, who asked him to become a tutor to his son Otto II.

In 972, Gerbert moved to Rheims and became a lecturer at the city's cathedral school. As noted earlier, the establishment of cathedral schools had been one of the pillars of innovation promoted by Alcuin in the field of education. For about seventeen years, Gerbert held his position at that school (with the exception of a period from 981 to 984, when he was appointed abbot of Bobbio and the subsequent short period he spent in Pavia), where he was able to fully develop his studies in various literary and scientific fields (arithmetic, geometry, astronomy, etc.) as well as write various works on mathematics and astronomy (*Liber de astrolabio, Libellus de numerorum divisione, De sphaera, De geometria, Regulae de numerorum abaci rationibus, Regula de abaco computi, Liber abaci*, and *Scholium ad Boethii arithmeticae institutionis*). His pupil Richerus, the author of a biography of Gerbert,[2] gives us fascinating details about his master's teaching of scientific subjects. Gerbert devoted particular attention and creativity to the scientific disciplines of the Quadrivium (rather than those of the Trivium, which were literary and philosophical in nature). In particular, Richerus describes the construction and methods of using the abacus, which we will discuss later, and the design of musical instruments, dwelling especially on the teaching of astronomy. For this teaching activity, Gerbert had a system of spheres and hemispheres with which he described the movement of the Sun, the Moon, the planets, and constellations. In a letter to his pupil Constantine of Fleury, the future abbot of Micy, Gerbert described in great detail the structure and use of one of these hemispheres, which was equipped with sighting tubes for astronomical observations.[3] Richerus calls this system of teaching

---

[2] Gerbert's biography is contained in Book III of the *Histories*; see R. di Saint Remi, *I quattro libri delle Storie (888–998)*, Pisa University Press, Pisa, 2008.

[3] See C. Sigismondi, La sfera di Gerberto, *Gerbertus*, 1, 2010.

'divine'. Its fame quickly spread to other schools. In those years, Gerbert also had intense epistolary exchanges with his pupils and other clergymen, from whom he frequently requested scientific books and manuscripts, such as when, for example, he requested ancient texts on astrology, rhetoric, and ophthalmics from the monk Rainardo of Bobbio or when he asked his friend Llobet of Barcelona for a treatise on astrolabe, in all probability the one written by the Persian Jew Masha'allah Ibn Athari.[4] His profound culture and mathematical skills even earned him the accusation of being a magician in league with the devil, which was maliciously made against him after his death.

At the same time, and especially after the period he spent at the cathedral school in Reims, Gerbert also carried out intense ecclesiastical, diplomatic, and political activities. He engaged both in supporting the succession of Otto III after the reign of his father Otto II and in supporting Hugh Capetus in his succession to the Frankish throne after the end of the Carolingian Dynasty. In 991, upon the death of Archbishop Adalberon of Rheims, Gerbert was elected to that position, but he encountered a level of hostility from both the king of France and the pope that went as far as excommunication. In 997, he was therefore forced to give up the office, taking refuge with Emperor Otto III and working to restore harmony with his opponents. The following year, the same Pope Gregory V who had forced him to renounce his role as archbishop of Rheims appointed him archbishop of Ravenna. In 999, by virtue of the powers that the Saxons emperors had acquired vis-à-vis the papacy, Gerbert was appointed pope by Otto III. The name Sylvester II that Gerbert adopted had a very ambitious meaning: Just as Sylvester I had flanked the Roman emperor Constantine, so Gerbert intended to flank Otto III in the noble endeavor of reconstituting the Western Roman Empire according to the inspiration of the *Renovatio Imperii Romanorum*. Gerbert died in 1003 after having been active in resolving conflicts within the Church, advocating the Cluniac reform and converting Slavic peoples.

---

[4] See A. Rucquoi, *Histoire médiévale de la péninsule ibérique*, Seuil, Paris, 1993, p. 146. The astronomer Masha'allah is believed to have participated in the founding of Baghdad by providing the caliph al-Mansur with positive astrological auspices.

Gerbert's mathematical works show a deep culture based on medieval classical texts (Isidore of Seville, Boethius, Bede). His most challenging work is *De geometria*, which begins by assigning third place to geometry in the Quadrivium after arithmetic and music, astronomy being the fourth discipline. The text consists of ninety-four chapters, about half of which essentially contain exercises and their solutions. Among them, we find some classic problems that had already appeared in Alcuin's text, which concern aspects of everyday life, with their respective methods of solution: how many trees can be planted in a plot of land of a given size, how many houses can be built within the walls of a round (or square) city of a given size, how many amphorae can be arranged in a well, how many tiles are needed to pave a basilica, and so on.[5]

Gerbert's interest in mathematics is also reflected in his prolific epistolary correspondence with pupils and other scholars. The letters with a scientific content that he sent to his pupil Constantine, abbot of Micy, are particularly interesting. In addition to the one we have already mentioned on the sphere for astronomical observations, we also recall a letter dedicated to algorithms for calculating multiplication and division, and another dedicated to *superparticular* numbers, a category of fractional numbers introduced by Boethius corresponding to ratios between two consecutive integers, for example, 3/2 (sesquipedale), 4/3 (sesquiterzo), and 5/4 (sesquiquarto).[6] In another letter sent to Adelboldo in 999 (on the eve of his ascension to the papal throne), he dwells on the method of calculating the area of an equilateral triangle if one knows the length of the side.

Without in any way detracting from the works mentioned earlier, the main reason why Gerbert plays a significant role in the history of calculus is related to the abacus named after him. As we have already stated, the abacus used in classical times consisted of a grooved rectangular wooden board, on top of which were engraved the letters corresponding to the

---

[5] N. Bubnov (ed.), *Gerberti Postea Silvestri II Papae Opera Mathematica (972–1003)*, Friedlander, Berlin, 1899.

[6] Superparticular numbers are important in music theory as they relate to the frequency relationships between notes; see P. Rossi, *Algoritmi matematici nelle lettere di Gerbert*, *Gerbertus*, 1, 19. 2010.

values of the Roman numbering system: I for units, X for tens, C for hundreds, and M for thousands. Pebbles (*calculi*) were placed in the grooves, representing the number of units, tens, hundreds, and so on. In the type of abacus invented, or in any case popularized, by Gerbert, instead of placing n pebbles ($1 \leq n \leq 9$) in a column, a token (called *apex*) was placed that carried the number n in Indo-Arabic numerals. Of course, the lack of a token in a column corresponded to zero.[7] To perform a calculation, one would arrange the digits of one of the operands at the top of the columns and the digits of the other at the bottom. Partial results (for instance, to perform multiplication) were always reached by using tokens in the rows below the first. There were twenty-seven columns of the abacus, three of which were devoted to decimals. The columns were grouped three by three, from right to left, three for the numbers from 1 to 999, three for the numbers from 1,000 to 999,000, and three for the millions, tens and hundreds of millions, etc., just as we separate the figures three by three with a comma. Each group of three columns was identified with an arc, which became known as the *Pythagorean arc*. The ways in which this abacus, which was also called the *mensa Pithagorae*, was to be used to perform the four arithmetic operations are presented by Gerbert himself in the text *Regulae de numerorum abaci rationibus*. Details on the use of Gerbert's abacus are also provided in the works of his pupils Richerus and Bernelinus.

Many have emphasized the innovative aspect of the use of tokens in Gerbert's abacus, pointing to him as the first European to have used Indo-Arabic numerals and algorithms to perform arithmetic operations. It is possible that Gerbert became acquainted with Indo-Arabic numerals during his stay in Spain through contact with Arab mathematicians,[8] but in truth, he uses Roman numerals exclusively in all his works, and the use of tokens with Indo-Arabic numerals is an exception. In fact, the question of

---

[7] An accurate description of Gerbert's abacus and the Indo-Arabic figures he used on the tokens can be found in the biography of Gerbert by N. M. Brown, *The Abacus and the Cross*, Basic Books, New York, 2010.

[8] Indeed, it seems that during his years in Spain, Gerbert went as far as the capital of the caliphate, Cordova. See M. Zuccato, Gerbert of Aurillac and a tenth century Jewish channel for the transmission of Arabic science to the West, *Speculum*, 80(3), 745, 2005.

I Ƭ Ƹ Ϋ Ϥ Ƅ ٦ 8 ٩

Figure 1.   Indo-Arabic numerals 1, 2, 3, 4, 5, 6, 7, 8, and 9 in ghubar characters.
Source: Redrawn from the *Codex Vigilanus*, 976 A.D.

who first used such numerals in Europe has aroused much discussion over the centuries. According to what we might call the 'Bible' of numerals, Georges Ifrah's text *Histoire universelle des chiffres*, the earliest European documents whose authenticity can be attested to in which the nine Indo-Arabic numerals appear come from northern Spain: They are the *Codex Vigilanus*, compiled in 976 in the monastery of Albelda de Iregua, in the Ebro valley, by a monk named Vigila, in which the nine digits appear in a commentary on Book III of the *Origines* of Isidore of Seville; and the *Codex Aemilianensis*, which was copied from the previous one in the nearby monastery of San Millán de la Cogolla in 992.

The monk Vigila states, "We should know that the Indians have the most subtle ingenuity and surpass all other peoples in arithmetic, geometry, and the other liberal arts. This is evident in the nine digits they use to denote all other numbers, of whatever order and magnitude. They have the following forms"[9] (see Figure 1).

The numerals are written in the *ghubar* characters used by Western Arabs. Indeed, it is well known that while the nine digits of Indo-Arabic origin used in Eastern Islamic countries are different from those that became established in Europe, they are derived from the variant used by Arabs in Western countries (Maghreb and Spain).[10] Other examples of tokens with the Indo-Arabic numerals *ghubar* appear in the 11th century in the works of Gerbert's student Bernelino.

The reason for the disputes about the first appearance of Indo-Arabic numerals in Europe seems to derive from the fact that, for centuries, the tokens with Indo-Arabic numerals, which were used to make calculations with the abacus more efficient, were called *Boethius' apices* because they appeared in the *Ars geometrica*. The thesis that Boethius knew the

---

[9] K. Devlin, *The Magic Numbers of Fibonacci*, Rizzoli, Milan, 2012, pp. 34–35.
[10] G. Ifrah, *Histoire universelle des chiffres*, Vol. II, Laffont, Paris, 1994, pp. 346–350.

Indo-Arabic numerals additionally led some nineteenth-century scholars to believe that these numerals had penetrated into Europe directly from India through the Neo-Pythagorean mathematical school of Alexandria, or even through contact that Pythagoras' own school had with India.[11] Today, however, we know that the *Ars geometrica* is the result of numerous later interpolations, if not even an apocryphal work compiled by an anonymous author in the 11th century and erroneously attributed to Boethius,[12] and so all the hypotheses developed in past centuries turn out to be completely unfounded.

In any case, the use of Indo-Arabic numerals in the 10th and 11th centuries appears to be entirely episodic, and one would have to wait until the first major translations of the works of Arab mathematicians in the 12th century for the numerals to be used in this way and for related algorithmics to consolidate and begin to spread.

The Translation of Arabic Texts in Spanish Monasteries and Cathedrals

Between the end of the 10th century and the 12th century, a series of events led to an impetuous cultural development in Europe, which the American historian Charles Homer Haskins summarized by the expression 'the Renaissance of the 12th century'. During this period, new cathedral schools were created, such as the Chartres school, which received a strong impetus from Bishop Fulbert (960–1028) — who was, if not a pupil of Gerbert, as claimed by some historians, nonetheless very close to him — and the first universities were born — the University of Bologna was founded in 1088 and the University of Paris dates from 1170.[13] Interest in the literary and philosophical culture of the Greek and Latin world became increasingly evident and led to an intensification of translations and reissues of ancient works. The famous expression "We are

---

[11] It is known, for example, that the Pythagoreans believed in metempsychosis, perhaps as a result of contact with the Brahmin culture.

[12] M. Folkerts, *"Boethius" Geometrie II. Ein Mathematisches Lehrbuch des Mittelalters*, Franz Steiner, Stuttgart, 1970.

[13] The university was formally recognized by the king in 1200 and by Pope Innocent III in 1215. The name *Sorbonne* was given later.

dwarfs on the shoulders of giants," which John of Salisbury attributes to Bernard of Chartres, whose pupil he was between 1114 and 1119, shows how it was clear to the most trained minds that in order to resume the development of letters and thought, it was necessary to start again from a thorough knowledge of classical culture. In fact, the above-quoted sentence continued thus: "We can see more things than they can and farther than they can, certainly not because of the acumen of our sight or the height of our bodies, but because we are lifted up and carried high by the stature of giants."

At the same time, on a different front, European culture was beginning to measure itself against Islamic culture. Relations between the Muslim world and the Christian world, which for two centuries after 1096 experienced the crisis of the Crusades, also had — in Sicily and Spain and more generally in the area of the western Mediterranean — phases of fruitful coexistence from both a commercial and cultural point of view. The Crusaders themselves, despite the rupture caused by the Crusades, brought back to Europe the absorbing experience of encountering a scientific culture that in those centuries was far more advanced than the European one. According to the historian of mathematics Carl Boyer, "The renaissance of knowledge in Latin Europe took place during the Crusades but, probably occurred in spite of the Crusades."[14]

The main region where the Latin translation of Arabic texts began to develop most intensively was Spain, or rather those territories of Spain that were gradually removed from Muslim rule by the struggle for the *reconquista.* There were many reasons for this. For one thing, as mentioned previously, the contact that existed during periods of peace between the two communities also favored cultural exchange. In addition, the return of Muslim possessions to Christian hands also made it possible to acquire important collections of Arab manuscripts, as happened, for example, with the reconquest of Toledo (1085) and Zaragoza (1118). This same territorial reconquest movement meant that Arabic-speaking minorities such as the *Mudejars* (Muslims) or *Mozarabes* (Christians) were welcomed into the population of Christian Spain and made important contributions to the translation and interpretation of Arabic texts. Finally,

---

[14] C. B. Boyer, *Storia della Matematica*, Mondadori, Milano, 1982, p. 293.

significant support for the translation movement came from Jews (or Jewish converts to Christianity) who, by virtue of the many contacts in the Mediterranean basin as a result of their business activities, were able to understand and use different languages.

During the years Gerbert had spent in Spain, the Christian area of the Iberian Peninsula only consisted of a strip in the north and included the county of Barcelona and the kingdoms of Aragon, Castille, Navarre, Asturias, and Galicia. The border with the caliphate of Córdoba was basically located along the Duero River and continued north of Zaragoza and then curved toward the coast by passing west of Barcelona, although we must keep in mind that the border was far from being consolidated at that time, and was subject to change during incursions by Muslims. It should be borne in mind that it was only in the last two decades of the 10th century that the Umayyad leader al-Mansur recaptured several localities (Coimbra, Salamanca, and Sepúlveda) and attacked and sacked Pamplona (978), Barcelona (985), Santiago (997), Burgos (1000), and several other localities.

In the Christian territories during this period, there was a flourishing of monastic initiatives. In addition to fulfilling a religious role, monasteries became cultural centers of great importance. Large archives of Arabic manuscripts were created in them, and translation and study activities were developed. We have already mentioned, for example, the monastery of Santa Maria de Ripoll, in the Pyrenees, at which Gerbert d'Aurillac had the opportunity to study mathematics and astronomy in depth during his stay in Catalonia. According to a catalog compiled in 1047, upon the death of Abbot Oliva, the library of Ripoll held 246 volumes, with works by Boethius, Macrobius, Aristotle, Cicero, Isidore, St. Augustine, Bede, and so on. Among them was an important collection of Latin texts on astronomy, mostly translations of Arabic texts.[15] The reputation of the monastery transcended borders and attracted scholars from various European regions.

The first major venues of the recovery and translation of Arabic texts, in the 10th and 11th centuries, were in particular the monasteries of the Ebro valley, which stretches from west to east at the foot of the Pyrenees.

---

[15] Rucquoi, 1993, *Op. cit.*, p. 145; see also Zuccato, 2005, *Op. cit.*, p. 749.

As we have noted, in the monasteries of Albelda (where there were many Mozarab monks and abbots) and San Millán de la Cogolla, activities were taking place not only on the reediting of Latin texts but also on the translation of Arabic texts. Greek works were also being rediscovered and translated into Latin from their Arabic version. At the end of the 12th century, another monastery in the Ebro valley, Rueda de Jalón, assumed an important cultural role. The last king of Zaragoza, Abd al-Malik ibn Hud, had taken refuge there in 1110 when the city was conquered by the Almoravids. On that occasion, Abd al-Malik had succeeded his late father and moved to Rueda, taking with him the important library of his grandfather Yusuf al-Mu'taman. After Zaragoza was conquered by the Christians in 1118, Abd al-Malik held out in the castle of Rueda until 1130, after which his stronghold was also conquered, at which point the library became part of the monastery's holdings.[16] Among the works believed to have been part of the library were Euclid's *Elements*, Ptolemy's *Almagest* and *Quadripartitum*, al-Khwarizmi's astronomical tables, and al-Battani's *Kitab al-zij al-sabi* (The Book of Astronomical Tables). Translations promoted by Bishop Michael of Tarazona were made by Hugh of Santalla, Hermann of Carinthia, Robert of Chester (or of Ketton), Plato of Tivoli, and others.[17]

During the 12th century, after the reconquest of Toledo by Alfonso VI of Castile, which occurred in 1085, and as a result of the expansion of Christian-occupied territories, which reached as far as the Tagus basin, the great cathedral library of Toledo became the most important Spanish center of translation of Arabic works. This activity was encouraged and sponsored by the successive archbishops of that period, notably the French Francis Raymond de Sauvetât (Raymond of Toledo, Ramon or Raimundo for the Spanish), who held office from 1126 until

---

[16]Yusuf al-Mu'taman ibn Hud was a valuable mathematician. He himself wrote the *Book of Perfection* (*Kitab al-Istikmal*), a compendium of the Greek mathematics of Euclid and Archimedes that also included contributions from ibn Qurra and al-Haytham.

[17]There are many doubts about the identity of some translators who worked in Spain. Their activity is attested by the notes they themselves made in their manuscripts, which often seem contradictory or inconsistent. For example, there is no agreement among scholars as to whether or not Robert of Ketton and Robert of Chester were the same person.

1151, and his successors John of Castelmoron (1151–1166) and Cerebruno (1166–1180).[18]

Many of the translators who worked in Toledo were Hispanic, such as Dominicus Gundisalvo (or Dominicus Gundissalinus, 1105–1181), archdeacon of Cuéllar, who is credited with the translation of more than twenty Arabic works, including several by Avicenna, al-Kindi, al-Ghazali, and al-Farabi, all philosophical in character, and who was himself a philosopher. Some of the translators came from territories under Muslim rule, such as John of Seville (Johannes Hispalensis), possibly a Mozarab who was active between 1118 and 1153. In addition to translating al-Khwarizmi's work on Indian numerals (under the Latin title *Alghoarismi de practica arismetricae*[19]), he translated works on philosophy, medicine, and alchemy. Some of the translators who played an important role were Jews, such as Yehuda ben Moshe, rabbi of the synagogue of Toledo, whose fame was so great that he became the personal physician of Alfonso X of Castile. Yehuda ben Moshe translated several works on geography, astronomy, and astrology.

Scholars from many regions of Europe flocked to Toledo. Among the most active translators in the 12th century, we might mention the Slavic Hermann of Carinthia (c. 1100–1160), who after training in the cathedral school of Chartres and traveling to Constantinople and Damascus devoted himself to translating Euclid's *Elements* and Ptolemy's *Planisphere*. Also worthy of remembrance are the Englishmen Daniel de Morley and Robert

---

[18] The translation activity carried out in the city of Toledo was perhaps overly mythologized in the 19th century. According to more recent studies, a real 'school of translation' in Toledo never existed, and the role of the city's archbishops was not as decisive as has been believed in the past; cf. R. Pergola, Ex arabico in latinum: Traduzioni scientifiche e traduttori nell'Occidente Medievale, *Studi di Glottodidattica*, 3, 74–105, 2009.

[19] The *Liber alghoarismi de practica arismetricae qui editus est a magistro Johanne Yspalensi* appears in the collection of manuscripts made by Boncompagni in the 19th century. The translation was dedicated to Raymond Archbishop of Toledo; see A. Genocchi, Bibliografia del Principe Boncompagni, *Annali di Scienze Matematiche e Fisiche* (Rome), VIII, 1857. The very important collection of medieval mathematical manuscripts made by Prince Boncompagni and their publication deserve to be better known and appreciated. Unfortunately, the collection was auctioned off and dismembered on the death of the prince.

of Chester (or of Ketton). In 1144, Robert of Chester translated a book on alchemy, the first to be known in Europe, and subsequently translated al-Khwarizmi's *Algebra* and his astronomic and trigonometric tables. Prominent among his translations is one of the Koran (*Lex Mahumet pseudoprophete*), which was commissioned by the Abbot of Cluny, Peter the Venerable, in 1142; he completed it in collaboration with Hermann of Carinzia, Peter of Toledo, Peter of Poitiers, and a Muslim who is only known by the name Mohammed. The Englishman Adelard of Bath (1080–1152) also stayed in Toledo during his wanderings that took him to Italy, Greece, and Syria. Adelard composed original scientific, mathematical, and astronomical works, but he was also one of the most important translators. He seems to have been credited with the most authoritative translation of Euclid's *Elements* and the translation of the astronomical tables and other works of al-Khwarizmi. Another noteworthy figure was the Scotsman Michael Scotus (1175–1232), who after a period in Toledo in the early 13th century, during which he devoted himself to translating works by Aristotle, Averroes, and Avicenna, moved to southern Italy, to the court of Frederick II, as an expert in mathematics, philosophy, and astrology.

The Italian Gherardo da Cremona (1114–1187) also worked in Toledo and was perhaps the most famous and prolific of the translators of the Toledo school. He was passionate about science and went to Toledo in search of Ptolemy's *Almagest* and stayed there, devoting himself entirely to translating scientific works from Arabic. According to the testimony of Daniel de Morley, Gherardo was helped in his translations by a Mozarab a certain Galippo, who used to translate Arabic texts into the 'Tholetan language' on the fly, texts that Gherardo then translated into Latin, a practice that seems to have been quite widespread. In addition to the *Almagest*, among the more than seventy works mentioned in the *Commemoratio librorum* compiled upon Gherardo's death by his colleagues (some twelve on astronomy, seventeen on mathematics and optics, twenty-four on medicine, etc.), Avicenna's *Canon of Medicine*, Euclid's *Elements* (from ibn Qurra's Arabic version), Aristotle's *Physics*, al-Farghani's *Elements of Astronomy*, Abu Bakr's *Liber mensuratorum*, and al-Khwarizmi's *Algebra* stand out.

The number of translators who worked in Spain, even outside the translation centers of the Ebro valley and Toledo, was extremely large.

Particular attention should be devoted to the aforementioned Plato of Tivoli (1110–1145), who also worked in Barcelona and translated mainly astronomical and mathematical works, and to various Jewish philosophers and mathematicians: Savasorda (Abraham ibn Hiyya, 1070–1136), himself the author of an important work on geometry that was translated into Latin by Plato of Tivoli, the *Liber embadorum*, which was certainly known to Leonardo Pisano, who was inspired by it in drafting the *Practica geometriae*[20]; Avendauth (Abraham ben Dawud, 1110–1180), the translator of Avicenna's *De anima*, whose name recurs in various manuscripts but whose identity is very uncertain; and Rabbi ben Ezra (Abraham ibn Ezra, 1092–1167), who is considered to be one of the most influential connoisseurs of scientific thought the Jewish community had cultivated in Muslim countries, a thought that he transmitted to the Jews of Western Europe.

Finally, we must take into account that in the case of some works, even important ones, the name of the translator is not known. This is the case with Abu Kamil's *Algebra*, versions of which remain in Arabic, Latin, and Hebrew. This work, which deals with the solution of first- and second-degree equations and indeterminate equations, was probably known to Leonardo Pisano, who was perhaps influenced by it when he wrote the *Liber abaci*.

As we have noted, and as already mentioned in Chapter 2, al-Khwarizmi's most important works were the subject of several translations. In particular, there is considerable uncertainty with regard to the translations of the book on arithmetic with Indian numbers (*Kitab al-hisab al-hindi*). No copy of the original remains. We have several manuscripts of this work that are traceable to four texts: *Dixit Algorizmi*, *Liber Ysagogarum Alchorismi*, *Liber alghoarismi de practica arismetricae*, and *Liber Pulueris*. The first, and probably the oldest, contains the most widely used version and appears to have been derived directly from the Arabic original, but its translator is unknown. The second is attributed to Adelard of Bath or Petrus Alfonsi because of the following incipit: "*Liber Ysagogarum Alchorismi in artem astronomicam a magistro A. compositus.*" The *Liber alghoarismi* is supposed to have been composed in Toledo in 1143, and as

---

[20] See Pergola, 2009, *Op. cit.*, p. 82.

mentioned earlier, seems to be attributable to Johannes Hispalensis, although this is disputed by some scholars.[21] Finally, the *Liber Pulueris* is probably an older and briefer version of the *Liber alghoarismi*.

Fortunately, we have somewhat more precise information regarding the translations of *Algebra*. There are more than fifteen extant manuscripts, which can be traced back to three different translations: one by Robert of Chester made in Segovia in 1145, one by Gherardo da Cremona made in Toledo in around 1170, entitled *Liber Maumeti filii Moysi Alchoarismi de Algebra et Almuchabala*, and a much later one made in the third quarter of the 13th century by William of Luni.[22]

---

[21] The historian André Allard believes that the *magister Johannes* who published the *Liber alghoarismi* in Toledo in 1143 (whom he calls John of Toledo) is not the same person as Johannes Hispalensis; see A. Allard, Les sources arithmétiques et le calcul indien dans le Liber abaci, in M. Morelli and M. Tangheroni (eds.), *Leonardo Fibonacci*, Pacini Editore, Ospedaletto (PI), 1994, p. 84.

[22] N. Ambrosetti, *L'eredità arabo-islamica nelle scienze e nelle arti del calcolo dell'Europa medievale*, LED, Milan, 2008, p. 126.

Chapter 5

Abacus Books and "Algorisms"

Early Abacus Books

Despite the translation of Arabic mathematical texts, the spread of computing methods based on Indo-Arabic numerals in Europe was very slow. As will be seen in the following, there would be no real impetus to use such methods until the 13th century, after the publication of Leonardo Pisano's seminal work *Liber abaci*, and especially in the 14th century due to the advantages they offered in the mercantile sphere. Before addressing this aspect, which would profoundly affect mathematical knowledge and the practice of calculus in Europe, we must note the fact that after the innovation of the use of the abacus introduced by Gerbert of Aurillac, so-called *libri d'abaco* (abacus books) began to appear. The purpose of these textbooks — whose roots were based on the earlier medieval literature related to *computus* and the practice of arithmetic — was to introduce young men who were being trained in the main cathedral schools of Western Europe to arithmetic.

At the same time as the works of Gerbert of Aurillac or a few years later, a number of contributions concerning arithmetic calculus appeared, including in particular work by the Benedictine monk Herigerus of Lobbes (925–1007), entitled *Regulae de numerorum abaci rationibus*, and work by Abbon of Fleury (ca. 950–1004), entitled *Commentarius in calculum Victorii* (which was also called *Tractatus de numero, pondere et mensura* with reference to the biblical verse "God ordered the world by measure, number, and weight"). Although these contributions may seem uninteresting because, as historians tell us, they are essentially tied to the tradition of Martianus Capella and Boethius, they highlight what the state of calculus was at the end of the first millennium; in any case, they are not devoid of innovative aspects. In particular, with regard to Abbon, it should be noted that his treatment of *computus* differs from earlier expositions. In fact, whereas in the 7th century, when *computus* became established in monastic culture, it was presented almost exclusively in textual terms (apart, of course, from the presentation of tables of the dates of Easter), and in Carolingian era diagrams began to be used to facilitate the understanding of concepts, in Abbon's presentation of *computus*, the algorithms are almost exclusively expressed with the help of tables and diagrams.[1]

Compared to these works, an undoubtedly more important role was played by the texts devoted to the use of the abacus that followed the groundbreaking work of Gerbert of Aurillac, beginning with *Liber abaci*, which was written in 999 by his pupil Bernelinus. In the 11th and 12th centuries, the production of works such as these spread further, and the first abacus schools emerged in which people were taught to perform arithmetic operations according to the methods popularized by Gerbert. Some important examples on the use of the Gerbertian abacus are the treatise on multiplication with the abacus written by the monk Hermannus Contractus in the mid-11th century, the treatise *De abaco* written by Rudolf of Laon in the early 12th century, and *Tractatus de abaco* by Gerlando of Besançon, which was also written in the 12th century. We have already mentioned Hermannus, who was a monk of the imperial abbey of Reichenau, on Lake Constance, concerning the *rithmomachia* game. He was called *contractus* — that is, crippled — because of a malformation caused by infantile paralysis. He was the author of several texts on mathematics and astronomy, including the treatise *Qualiter multiplicationes fiant in abaco*, *Regulae in computum*, and the treatise *De geometria*, as well as two texts on the astrolabe. He also wrote histories and chronicles of Swabia — about the emperors Conrad and Henry — and a history of the world from the birth of Christ to his day.

---

[1] Immo Warntjes, Introduction: State of Research on Late Antique and Early Medieval Computus, in Immo Warntjes and Dáibhí Ó. Cróinín (eds.), *Late Antique Calendrical Thought and its Reception in the Early Middle Ages*, Brepols, Turnhout, 2017, p. 27.

Gerlando[2] was also a religious figure, the canon of St. Paul's Abbey in Besançon. In addition to the *Tractatus de abaco*, of which several manuscripts still remain — which indicates that it was widely distributed — he wrote *Regula de dialectica, Tractatus de grammatica*, a collection of astronomical tables, and a treatise on ecclesiastical computation called *Computus Gerlandi*, which is often cited as a work carried out "in imitation of Bede" because of its affinity with the English monk's writings. His expertise is attested to by the expression that appears in a dedication: "*Gerlando scientia trivii quadriviique onerato et honorato*" (Gerlando, who bore the burden and the honors of the science of trivium and quadrivium). In another document, moreover, it is reported that when Adalberone, the bishop of Trier, went to the Diet of Frankfurt in 1148, "He led Master Gerlando of Besançon and Master Theodoric of Chartres with him, and during this voyage made by them sailing on the Rhine, they engaged in a dispute with full satisfaction of this prelate."[3] In his treatise on the abacus, Gerlando illustrates how to perform operations of multiplication and division, following the methods outlined by Gerbert, and explicitly refers to the use of "tokens" on which nine characters were shown (but without saying that they were Indo-Arabic characters).

Also datable to the first half of the 12th century is the text *Regulae abaci* written by Adelard of Bath, a philosopher, naturalist, mathematician, and astrologer, a figure we have already come across as one of the most influential translators of Arabic texts, and himself the author of important original works such as *Questiones naturales* and discussions on human nature and the topics of astronomy, botany, and zoology. Adelard

---

[2] Historiography does not agree on the identity of Gerlando. Various sources attest to the existence of a Gerlando (sometimes called Gerlando Computista) who was born around 1015 in Besançon and died in 1100 in Agrigento, the city where he had become bishop and of which he is considered the patron saint; there was also a Gerlando, canon of St. Paul' Abbey in Besançon, who lived in the 12th century and died after 1148. The historian Nadia Ambrosetti identifies the former as the author of the text *Abacus libellus de scientia multiplicandi*, which would have been written around 1080; see Ambrosetti, *L'eredità Arabo-Islamica*, p. 100. Other authors identify the author of *Tractatus de abaco* as the latter, based on coeval documentation.

[3] See Anonymous, *Intorno al* Tractatus de abaco *di Gerlando*, in *Bullettino di bibliografia e di storia delle scienze matematiche e fisiche*, tome X, Rome, 1877, p. 654.

left England in the late 11th century and studied the arts of the Trivium and Quadrivium at the cathedral school of Tours. In 1113, he went to teach in Laon, the site of an important school of theology. He then began a series of journeys that lasted seven years and, as we have seen, also took him to Toledo. Finally, he returned to Bath in 1122. *Regulae* was written in the first decade of the century, thus before Adelard had contact with Arabic culture.

In England at approximately the same time, around 1115, another mathematician, Turchillus, known as a *computista*, drafted another work, *Regunculae super abacum*, in which an introduction to the abacus is provided with reference to the nine digits, saying that they come from the Pythagoreans — but also that their names come from the Arabs — and the rules for calculating multiplication and division are explained. In addition, a particularly novel aspect, the practical principles of administrative accounting, is introduced regarding the management of lists relating to give and take. A typical example of a calculation provided by Turchillus is as follows:

> Suppose there are 23 knights and each owes you 6 marks. You want to know what is the total number of marks that you get from these small sums, that is, 23 times 6 marks.

Another example (or exercise) deals with the division of a sum among 288 workers and yet another deals with the surface area of Essex, which according to William the Conqueror's accounting was estimated to be 2,650 *hides* (a *hide* was a measurement of area corresponding to about 120 acres).[4]

One characteristic feature of the abacus treatises of that period is that they frequently re-proposed texts by previous authors. Three 12th-century manuscripts from the Boncompagni collection presented in an article published in 1877[5] are illuminating in this respect. The first manuscript

---

[4] Reginald Lane Poole, *The Exchequer in the Twelfth Century*, Oxford, 1912, pp. 48–50.

[5] The article reports only excerpts from the three manuscripts; see Peter Treutelein, Intorno ad alcuni scritti inediti relativi al calcolo dell'abaco, in *Bullettino di bibliografia e di storia delle scienze matematiche e fisiche*, volume X, Rome, 1877, pp. 589–594.

contains Bernelinus' *Liber abaci*, Gerlando's *Regulae cuiusdam super abacum*, and Hermanni Suedi's *Regulae in computum*, rules for computing division and multiplication, and writings on music and astronomy. The second contains Bernelinus' text, *Liber de utilitatibus astrolabii* by Hermannus Contractus, and parts of Gerlando's text. The third contains Hermannus' *Liber de utilitatibus astrolabii* and *Libellus de mensura astrolabii,* as well as his rules for multiplication and division on the abacus. It can be seen that in these texts, the description of the algorithms of multiplication and division is still presented in a style that might be called "rhetorical" because everything is described in verbal terms with examples that only make use of Roman numerals. The nine "marks" (which are also called *insignes caracteres* in one of the manuscripts) were evidently only used for performing calculations on the columns of the Gerbertian abacus and not for representing data and results, for which the Roman representation was still used.

In the next few pages, we will see how the establishment of the Islamic mathematical culture and at the same time the demands of efficiency required by mercantile calculation brought about a fundamental shift in abacus books. While the Gerbertian abacus retained the duplicity of the ancient instrument and the innovative methods of calculation based on Indian numerals, with the work by Leonardo Pisano, the disappearance of the instrument was definitively sanctioned, and the abacus books became exclusively an instructional guide for arithmetic calculation or, more generally, mathematics textbooks.

Leonardo Pisano

The 12th and 13th centuries marked an impressive economic, social, and cultural awakening throughout Europe, particularly in Italy. The markets that opened up in the Levant as a result of the Crusades made a decisive contribution to this process, but, and perhaps above all, it was also because of the ideas and innovations that arrived along with goods from the eastern Mediterranean basin and penetrated Europe through Spain, Southern France (Languedoc and Provence), and Italy. This evolution affected not only the primary goods of human existence such as food and housing but also the means of production. We may recall, for example, the invention of the moldboard, which enables the plow not only to cleave the soil but also to turn over the clods, and other innovations that made agricultural work more productive. At the same time, the rebirth of cities contributed to the development of a merchant class and resulted in the need for them and city administrators to manage accounts efficiently.

It was in this context that the man who is rightly considered to be the initiator of the revival of European mathematics, Leonardo Pisano,[6] the most creative and influential mathematician who lived in Western Europe after Diophantus and before Pierre de Fermat,[7] was born and worked in Italy. He should not only be credited with the dissemination in Italy, and in the mercantile environment of Western Europe more generally, of algorithmic techniques based on the positional numerical system of Indo-Arabic origin but above all with the creation of an extraordinary wealth of notions and mathematical results of the greatest importance to algebraic calculus and its applications.

Leonardo's date of birth is unknown,[8] but it is believed that he was born between 1170 and 1180. His father Guglielmo dei Bonacci was a merchant. At a time we can date as around 1185, he took up a post at the customs house in Bugia (in Arabic Béjaïa) on the coast of present-day Algeria as *publicus scriba pro Pisanis mercatoribus*. Essentially, this involved performing a notarial task and representing Pisan merchants before the local authorities.[9]

Pisa had achieved a very prominent position in mercantile activity between the 10th and 12th centuries and, with Genoa, had become the leading maritime power in the western Mediterranean. In 1034 and 1087, Pisan forces attacked and sacked cities in North Africa (Bona and Mahdia), and in 1063, they played an important role in the reconquest

---

[6] Today, Leonardo Pisano is better known by the name Fibonacci, meaning a descendant of the Bonacci family. In fact, this name was only coined in 1838 by the historian Guillaume Libri.

[7] Leonardo Pisano, *The Book of Squares*, introduction and comments by Ettore Picutti, in *Physis*, vol. XXI, 1979, p. 198.

[8] In some documents, Leonardo Pisano is called *Bigollo* ("Discreet and wise master Leonardo Bigollo"), but most scholars believe it is a nickname that meant "traveler".

[9] Marco Tangheroni, Fibonacci, Pisa e il Mediterraneo, in Morelli and Tangheroni (eds.), *Leonardo Fibonacci*, pp. 16–17.

(and pillage) of Palermo, which was under Muslim rule. Later on, peaceful commercial activity took over. Pisan *funduks* (warehouses) were installed in various cities on the African coast and beginning in the 12th century in the Holy Land, as part of the Latin kingdoms founded during the Crusades, particularly in Acre, Tyre, and Jaffa. Trading with African coastal cities in the western Mediterranean became particularly intense. The first trade treaty between Pisa and Bugia dates from 1133 and was renewed for twenty years and extended to other Maghreb cities (Almería, Ceuta, Oran, and Tunis) in 1186 (precisely during the period when Leonardo's father was in Bugia) and 1211. Pisa imported grain, leather, hides, alum, wax, and gold from Bugia, and exported weapons, timber, and various manufactured goods there.[10]

In the course of their trading and bargaining activities, the Pisan merchants had probably grasped the greater efficiency the Arabs had achieved in handling accounting with computational techniques based on the positional system. In the preface to his most important work, *Liber abaci*, Leonardo Pisano relates how his father took him to Bugia in his youth for the very purpose of educating him in the study of mathematics (*abbaci*) so as to guarantee him a future of prosperity (*inspecta utilitate et commoditate futura*). As he himself says, "There by the wonderful instruction in the art of the nine digits of the Indians, to be introduced and to know such art pleased me so much above all things." Then, since it can be assumed that Leonardo had also had a basic education in letters and calculus in Pisa in his childhood, his father's decision to have him learn the methods of calculation from the Arabs leads one to think that, according to the mentality of the merchants of the time, the importance of learning such methods in order to competently and efficiently carry out mercantile activities was already clear.

After his time in Bugia, Leonardo made a series of trips that took him to Egypt, Syria, Greece, Sicily, and Provence. Based on the works he wrote later on, we understand that in Bugia and during his travels to Islamic countries, where he also worked as a merchant,[11] he became

---

[10] *Ibid.*, pp. 24–25.

[11] Leonardo calls the places he visited in various regions of the Mediterranean *loca negotiationis*. His presence in Constantinople in around 1200, where he carried out "the task of

familiar with Arabic mathematics and also with Greek mathematics (particularly the works of Euclid and Diophantus), probably from Arabic versions or translations from Arabic into Latin that were already circulating and which he had the opportunity to read. Again, in the preface to *Liber abaci*, Leonardo says, "On the whole the algorithm and even the Pythagorean arc appeared to me almost a mistake when compared to the Indian methods." By these words, Leonardo meant to affirm the superiority of Indo-Arabic methods of calculation over the use of the abacus (*arcus Pictagore*) and the related methods of calculation developed by Gerbert (*algorismus*).[12]

Back in Pisa, Leonardo put his skills to good use by devoting several years to writing a number of mathematical works that were to become milestones in the history of European mathematics and were among the main references until the 16th century. In 1202, we have the first version of *Liber abaci*, which he sometimes cites as *Liber maior de numero*,[13] and which was later republished after a revision in 1228, which we will discuss at length in the following. In 1220 (or 1221), he published *Practica geometriae*, in which, after an introduction containing definitions taken from Euclid's *Elements*, there are eight sections devoted to calculating the surfaces of plane figures, calculating volumes of solid figures, extracting square and cubic roots, and solving other problems of a geometric nature, some of which were inspired by Plato of Tivoli's *Liber embadorum*.[14] This work was followed in 1225 by *Liber quadratorum*. This text contains

---

an interpreter for the Pisan commercial outpost and where he could have met local mathematicians" has been attested to; see Ambrosetti, *op cit.*, p. 219. This seems to be confirmed by the fact that in various parts of *Liber abaci*, Leonardo mentions problems that were proposed to him by "masters" in Constantinople.

[12] Laurence E. Sigler, *Fibonacci's Liber abaci*, Springer, Berlin, 2003, pp. 15–16. See also Ambrosetti, *op cit.*, p. 219.

[13] Another text entitled *Liber minor de numero* (or *Book of lesser guise*) is probably lost, as is a commentary on Book X of Euclid's *Elements* of which there is a trace in Leonardo's writings; see Veronica Gavagna, *Leonardo Fibonacci*, in Antonio Clericuzio and Saverio Ricci (eds.), *Enciclopedia italiana di scienze, lettere ed arti. Il contributo italiano alla storia del pensiero-Scienze*, Istituto della Enciclopedia Italiana, Rome, 2013, pp. 192–195.

[14] Devlin, *I numeri magici di Fibonacci*, p. 77.

twenty propositions concerning the properties of squares, set forth in a thorough manner using the correspondence between lengths of segments and numbers and accompanied by demonstrations that, while still written in a rhetoric style, are logically valid and can be appreciated as excellent mathematics.[15] Also, at an unknown date, Leonardo drafted a florilegium of arithmetical and geometrical problems (*Flos Leonardi Bigolli Pisani super solutionibus quarundam questionibus ad numerum et ad geometriam, vel ad utrumque pertinentium*), a short text containing altogether fifteen algebraic problems on determinate or indeterminate equations of the first degree. Finally, also at an unknown date, Leonardo wrote an *Epistle* in which he sets forth various problems of indeterminate analysis, of the same type as the so-called "bird problems" we encountered in the work of Abu Kamil.

The meeting Leonardo had in Pisa with Frederick II of Swabia, Emperor of the Holy Roman Empire from 1220 to 1250, was particularly important for the later development of his work. Frederick II was a lover of culture and science (he founded the University of Naples in 1224) and spoke six languages: Latin, German, French, Greek, Arabic, and Sicilian. He loved to surround himself with learned and knowledgeable Latin, Arabic, Greek, and Jewish scholars, some of whom accompanied him to the Holy Land in 1228 on the occasion of the Sixth Crusade and engaged in confrontations and debates with Arab scholars at the court of Saladin's nephew, the Ayyubid sultan al-Malik al-Kamil. According to some sources, Leonardo's encounter with Frederick II and his court of scholars occurred in 1220,[16] and it was recorded by Leonardo himself in some of his works. Leonardo was introduced to the court by Master Dominicus, perhaps Dominicus Hispanus, one of the translators who had worked in Toledo and then moved to the imperial court. To Master Dominicus Leonardo dedicated the *Practica geometriae* From then on, he came into close and frequent contact with other wise men in Frederick's sphere. In

---

[15] Leonardo de Pisa, *El libro de los números cuadrados*, introduction by Paul Ver Eecke, EUDEBA, Buenos Aires, 1973.

[16] According to other scholars, the meeting took place in Pisa in 1226, but in that case, it must be assumed that Leonardo's contact with the scholars of Frederick's court, or with Frederick himself, had begun earlier.

particular, the second edition of *Liber abaci* is dedicated to Michael Scotus, whom we have already mentioned as a mathematician and philosopher, and who came to Frederick's court in 1227. Leonardo claimed that he drafted the second version with corrections on the basis of the criticism Scotus had directed at him, following an in-depth examination of the work. The genesis of *Liber quadratorum* is interesting. According to what the author wrote in the prologue of his dedication to the emperor,[17] the text was written because of encouragement he received from another of the learned men of Frederick's court, John of Palermo, who had challenged him to find a square for which adding 5 or subtracting 5 still yielded squares (a result that obviously has no solution on the integers, but which Leonardo found on the rationals).[18] Two other questions posed by John of Palermo are treated in *Flos*. Finally, *Epistola ad magistrum Theodorum* indicates that Leonardo was also in contact with Theodore of Antioch, an astrologer of probable Syrian origin who was sent by the Sultan of Egypt to Frederick's court.

After years of more intense scientific production, it appears that Leonardo collaborated with the municipality of Pisa, teaching accounting techniques and administrative management. In fact, the last available document, which is dated 1241, attests to the payment of a fee with the following words: "Let him be assigned twenty lire by the municipality and by the public treasury, by way of annual wage or salary, in addition to the usual benefits, and furthermore let it be that the same [Leonardo] serves the municipality of Pisa and its officials as usual in the practices of the abacus."[19] As well as Leonardo's date of birth, the date of his death is also unknown. It was certainly after 1241 and can be assumed to be between 1242 and 1250.

---

[17] In the prologue to *Liber quadratorum*, Leonardo makes another important statement regarding Frederick II's interest in the mathematical sciences: "Understanding from those who return from the imperial court that your sublime majesty delights in reading that book I have composed of numbers [the *Liber abaci*] and that she delights in sometimes hearing subtleties of geometry and numbers."

[18] The number $11 + 2/3 + 1/144$, square of $3 + 1/4 + 1/6$, is a solution; in fact, adding and subtracting 5 gives the squares of 4.08333... and 2.58333....

[19] *Ipseque pisano Comuni et eius officialibus in abbacatione de cetero more solito serviat*; see Ambrosetti, *op cit.*, p. 220.

Fibonacci's work was a milestone in the development of European mathematics. Certainly, the contents of his work were influenced by the mathematical knowledge of the time, and in particular by the works produced by Islamic culture, but as some scholars have said, the greatness and importance of Leonardo Pisano's work is not measured in the originality of its contents but in the completeness and organic nature of the way in which it was presented.[20] All the old mathematical knowledge is assumed and critically discussed. Fibonacci associates mastery of the subject with a taste for inquiry as he claims: "Many proposed questions come to blossom — hence the title *Flos* of one of his most original works — generating innumerable other questions, just as small plants spring forth from roots." Another fundamental aspect is the formal approach of Leonardo's work (while still within an essentially rhetorical style of exposition), namely, the fact that after centuries in which the formulation of results had been favored over demonstration, he took up the Greek tradition of providing deductive logical support for the mathematical propositions he presented. Lastly, it seems important to underline that, as we shall see more clearly by analyzing *Liber abaci*, in addition to presenting algorithms of arithmetic calculation through numerical examples, Fibonacci presented them in an abstract way on many occasions: that is, by describing computational steps, albeit in a rhetorical style, in a data-independent way.

From a practical point of view, the most important aspect of Fibonacci's work is the impact his works, particularly *Liber abaci*, had on the everyday use of calculus. Despite great resistance from those who feared losing the power they derived from skill in using the abacus or those who feared that the new methods of calculation based on Indo-Arabic numerals might facilitate cheating (consider that in 1299, in Florence, the articles of the *Arte del Cambio* still forbade its members from keeping accounts using Indo-Arabic numerals), the techniques of arithmetic were no longer the patrimony of scholars in convents or a few other connoisseurs, but spread to the mercantile world and public administrations.

---

[20] Antonio Thiery, *Federico II e le Scienze*, in Angiola Maria Romanini (ed.), *Problemi di metodo per la lettura dell'arte Federiciana. Federico II e l'arte del Duecento italiano*, vol. II, Congedo Editore, Galatina, LE, 1980.

Liber Abaci: **Mathematics**

As noted earlier, *Liber abaci* was of great importance in the establishment and dissemination of computation techniques based on Indo-Arabic numerals and remained a staple in the education of scholars and merchants in Western Europe for many years. It deserves closer examination because it marked a significant turning point in the field of mathematical publications and was an important milestone in the history of algorithms, especially regarding the role computational techniques played in mercantile applications and in everyday life more generally. In fact, as we have already made clear, beginning with the introduction of the Gerbertian abacus, books on the subject began to focus on the execution of the algorithms realized by manipulating the tokens with Indo-Arabic digits rather than on the way the operations were performed according to the traditionl use of the instrument. In his work, Fibonacci accomplished a further step in this direction, and in spite of the title, the abacus tool disappeared for good. The whole work is devoted to the presentation of algorithms: On the one hand, the main computational algorithms for arithmetic operations and resolving arithmetic and algebraic (and to a small extent geometric) problems are described in a flat but technically sound manner and, on the other, a great wealth of computational applications in the most varied fields of mercantile interest are provided.

The book has fifteen chapters. The first seven are devoted to four arithmetic operations performed on integer numbers and fractional numbers. Chapters eight through eleven are devoted to applied problems (buying and selling goods, conversions between units of measurement, barter, an analysis of investments in the establishment of mercantile companies, the exchange of coins, etc.). Chapter 12 is dedicated to solving first-degree equations using the false position technique and a variety of other problems (including the rabbit problem that gave rise to the well-known Fibonacci succession). Chapter 13 is devoted to solving first-degree equations using double false positions. Finally, the last two chapters present a series of geometric and algebraic problems (the calculation of square and cubic roots, the calculation of volumes of various geometric figures, solutions to second-degree equations, etc.) that were very much inspired by the works of Euclid and al-Khwarizmi.

Let us look in more detail at the contents of the chapters of a mathematical character, which will give us a way to understand the more particular aspects of Leonardo's proposed algorithms, as well as the techniques of what he calls "the art of computing." We will then examine the contents of the chapters with an applicative character.

The first chapter, which follows the dedication to Michael Scotus and the autobiographical prologue mentioned earlier, mainly introduces the methods of representing numbers. As if to mark the fundamental change brought about by the use of Indo-Arabic numerals, the chapter opens with the following sentence:

"The nine Indian digits are: 9 8 7 6 5 4 3 2 1.[21] With these nine digits and the 0 sign, which the Arabs call *zephyr,* any number can be written, as we demonstrate below." In particular, it should be noted that for the first time, 0 is mentioned explicitly as one of the characters that allow the representation of any number. Throughout the book, integer or fractional numbers are always written with Indo-Arabic numerals, and Roman characters are only used exceptionally (usually in a discursive context). In his representation of numbers with many digits, Leonardo also recalls the desirability of grouping them three by three, a reminder of the arc in the Gerbertian abacus. In any case, however, it is interesting to note that, in the first chapter, Leonardo Pisano also gives space to another tool that he believes is necessary for computation, the digital calculus, which he refers to as "a wise invention of the ancient." Leonardo's presentation of the digital calculus, which is supported by an illustration, is quite similar to that offered by Bede some five hundred years earlier. Leonardo states, "Those who wish to know the art of calculation, its subtleties and ingenuity, must know how to calculate with digital figures." The fact that the digital calculus is still important in the art of computation appears clearly in the following chapter, in which the algorithm of multiplication is illustrated and Pisano explains how carryovers and partial results can be stored

---

[21] The nine digits are presented — counterintuitively — in the order 9...1, and not 1...9 (not only here but also in all previous writings in which Indo-Arabic digits were introduced) due to the fact that Arabic texts are written from right to left and not from left to right as is customary in Europe. Note also that Fibonacci defines the nine digits as Indian and the zero as Arabic.

(*serventur in manu*): that is, memorized with digital representation. The first chapter closes with the presentation of the last tool needed to perform complex calculations: elementary addition and multiplication tables.

The second and third chapters are devoted to addition and multiplication.[22] While addition is no different from the method used today, except for the habit of placing the operands at the bottom and the result at the top (as opposed to what we do today), in the case of multiplication, two different methods are proposed that are innovative compared to those that had been used before. In the earlier tradition, as we have had occasion to observe, multiplication was based either on the ancient technique of duplication and halving or on the technique of gradually erasing one of the operands (on the powder abacus) by progressively adding the result of the partial products to it. Fibonacci marks a discontinuity by reconnecting directly with the Arabic sources he had occasion to know. The first method he proposes is found in all four Latin versions of al-Khwarizmi's text[23] and derives from the interpretation of a number as a polynomial in the base 10. For example, if we consider the two numbers 1,234 and 5,678, we have

$$1,234 = 1 \times 10^3 + 2 \times 10^2 + 3 \times 10 + 4$$

$$5,678 = 5 \times 10^3 + 6 \times 10^2 + 7 \times 10 + 8$$

and so the product of the two numbers corresponds to the product of two polynomials:

$$\begin{aligned}
1,234 \times 5,678 = {}& 1 \times 5 \times 10^6 + (1 \times 6 + 2 \times 5) \times 10^5 + (1 \times 7 + 5 \times 3 \\
& + 2 \times 6) \times 10^4 + (1 \times 8 + 4 \times 5 + 2 \times 7 + 6 \times 3) \times 10^3 \\
& + (2 \times 8 + 6 \times 4 + 3 \times 7) \times 10^2 + (3 \times 8 + 4 \times 7) \times 10 \\
& + 4 \times 8 = 5 \times 10^6 + 16 \times 10^5 + 34 \times 10^4 + 60 \times 10^3 \\
& + 61 \times 10^2 + 52 \times 10 + 4 \times 8 = 7 \times 10^6 + 0 \times 10^5 \\
& + 0 \times 10^4 + 6 \times 10^3 + 6 \times 10^2 + 5 \times 10 + 2 = 7,006,652.
\end{aligned}$$

----

[22] For the first time, operations are carried out "in columns", as they are taught in elementary schools today, an arrangement of operands that was clearly not feasible with Roman numerals.

[23] Cf. Allard, *Les sources arithmétiques*, p. 91. It should be noted that this method also appears in Indian sources in which it is traced back to the third sutra; see Jagadguru Swami Sri Bharati Krsna Tirthaji Maharaja, *Vedic Mathematics*, pp. 40–48.

The problematic aspect of a multiplication algorithm such as this (which is presented by Leonardo several times, both abstractly and with reference to specific numerical examples, as the number of digits of the two factors varies) is that it requires one to keep in mind (or in hand, as Fibonacci dictates) several intermediate results in addition to carryovers.

Leonardo presents the second algorithm for multiplication at the beginning of the third chapter. It too derives from Arabic sources, specifically from the mathematician al-Uqlidisi (10th century).[24] The method, which Leonardo describes as "very much appreciated," consists in arranging in a checkerboard fashion (*in forma scacherii*) the result of multiplying the individual digits of the second factor (written vertically to the right of the checkerboard from bottom to top) by the first factor (written horizontally in the top row from left to right). The first row of the table contains the final result, which is obtained by diagonally summing the numbers contained in the rows below. Suppose we need to multiply the numbers 4,321 × 567. The intermediate and final data and results are organized as shown in Figure 1.

The final result is obtained by summing the partial results diagonally from left to right: 7, 4 + 6, (1) + 2 + 2 + 5, and so on. The success of the algorithm was enshrined in the following centuries by Arab, Byzantine, and Western mathematicians. It is not difficult to recognize in it the multiplication algorithm we use, despite (insignificant) differences in the placement of the two factors and the result and in the stacking of the partial products.

The subsequent Chapters 4 through 7 complete the exposition of arithmetic operations, including subtraction and division, for which the methods presented are substantially similar to those we use (again,

2	4	5	0	0	0	7	
			4	3	2	1	
		3	0	2	4	7	7
		2	5	9	2	6	6
		2	1	6	0	5	5

Figure 1. "Checkerboard" multiplication table of 4,321 × 567.

---

[24] See Allard, *op cit.*, p. 90.

minus details of the graphical setting). The treatment of division gives Leonardo the opportunity to introduce prime numbers (which, he says, are called *hasam* by the Arabs). Fibonacci calls the compound numbers "regular" and the prime numbers "irregular". In his discussion on division, Leonardo presents many examples of division with respect to prime numbers (17, 19, etc.), suggesting that where the divisor is a compound number, it should be broken down into prime factors and then the dividend should be divided away by such different factors. Even in prime factor decomposition, Leonardo shows his algorithmic intuition, as he observes that to verify whether an odd number n is a prime number, it is sufficient to test whether it is divisible by odd numbers less than or equal to $\sqrt{n}$.

The treatment is extended from integers to fractional numbers for all the arithmetic operations considered. The use of fractions in Fibonacci's work requires particular attention. Leonardo introduces various types of fractions: simple ones — that is, with a numerator (*denominatus*) and a denominator (*denominans*), like those to which we are accustomed[25] — and compound ones, with more numbers in the numerator and as many numbers in the denominator.[26] For simple fractions, Leonardo also presents an algorithm called "universal" for the decomposition (separation) of a generic fraction into unit fractions (called "Egyptian fractions") and also presents seven different methods of simplification — that is, reduction to the lowest terms.

Leonardo presents three different types of compound fractions (also stemming from his knowledge of Arabic mathematics). In each case, fractions can be followed by an integer to represent an integer value plus the value of the fraction. Without going into the merits of all three types (in part, in view of the fact that these compound fractions were difficult to manage and gradually disappeared from use), we simply consider the first type, which offers some interesting aspects. We can illustrate it with the following example:

---

[25] Leonardo calls the fractional numbers *fracti*, meaning "broken", and the fraction sign *virgula*, meaning "stick".

[26] Compound fractions are also called "threaded" fractions or fractions "in degrees" in the literature.

The fractional number

$$\frac{2\ 3\ 4\ 7}{4\ 5\ 3\ 2}\ 6$$

corresponds to: $2/(4 \times 5 \times 3 \times 2) + 3/(5 \times 3 \times 2) + 4/(3 \times 2) + 7/2 + 6$.

Calculating the value of these fractions is laborious, but one of the most interesting features is that a direct correspondence can be established between these fractions and the decimal representation of numbers. For example, the fractional number

$$\frac{5\ \ 1\ \ \ 4\ \ \ 1}{10\ 10\ 10\ 10}\ 3$$

corresponds to the number: 3.1415.

In actual fact, Leonardo rarely uses decimal numbers. The mathematician Laurence Sigler, who edited the English translation of *Liber abaci*, notes that this may be motivated by the fact that in Fibonacci's world, decimals were little used and units of measurement were often not divided into tenths but into different fractions (for example, the *lira* was divided into 20 *soldi* and *soldi* into 12 *denarii*); however, it is interesting to note that the compound fractions used by Leonardo brilliantly allowed for the expression of these subdivisions as well.

For example, if we want to express the sum of 6 *lire*, 2 *soldi*, and 3 *denari*, we can simply use the fractional number

$$\frac{3\ \ 2}{12\ 20}\ 6$$

which corresponds precisely to *lire* $6 + 2/20 + 3/(20 \times 12)$.

In addition, the compound fraction can also be seen as a synthetic representation of a series of cascading simple fractions. For example, the fraction

$$\frac{2\ 3\ 4\ 7}{4\ 5\ 3\ 2}$$

corresponds to the simple fraction $(7 + (4 + (3 + 2/4)/5)/3)/2$.

Similar properties apply to the other two variants of compound fractions, which Fibonacci represents with the fraction symbols o— and —o, but which, as noted earlier, we will not discuss here.

Other peculiar aspects of *Liber abaci*'s chapters on arithmetic concern the use of the proof of nine to verify the correctness of the four operations. The proof of nine was known earlier, but what merits being noted is that Leonardo Pisano not only uses it for the four operations and for multiplication, as had been the case previously, but also extends this proof method by using residues with respect to other numbers (7, 11, 13, etc.), thus showing non-trivial skill in the use of modular arithmetic. In conclusion, a final interesting aspect of these chapters concerns the methods of exposition of algorithms, which are mainly given by means of examples, but sometimes, as in the case of the algorithm of the greatest common divisor (an algorithm that the author rightly traces back to Euclid), are presented abstractly in a rhetorical style and without the use of symbols or abbreviations. In many other cases, conversely, demonstrations or illustrations of algorithms are always proposed by a rhetoric method, but with the help of abstract notation. This is the case, for example, when the validity of the proof of nine is discussed.

As mentioned earlier, Chapters 8 through 12 present various applications (which we will discuss in the next section), but they are nevertheless also rich in mathematical content of various kinds. In particular, the mathematical tools used to solve the problems in Chapters 8 through 11 are proportions, which are basically addressed with what are known as the rules of "simple three" (direct and inverse) and "compound three". In Chapter 8, proportions are used to address problems of buying and selling goods, in Chapter 9 they are used for problems of bartering between goods and in the exchange of metals for coins, in Chapter 10 they are used to determine the division of the profits of a company given the shares held by the partners, and in Chapter 11 they are used to determine the value of coins minted with different alloys. For these chapters, Leonardo claims to have been inspired by an Arabic author, Ametus, probably Ahmad ibn Yusuf al-Baghdadi, who lived between 835 and 912 and was the author of *Book of Proportions*, which was translated into Latin by Gherardo da Cremona.

Chapter 12 is highly composite and consists of nine parts in which very different mathematical techniques are used: continuous proportions, sums of arithmetic series, sums of geometric series, solution of first-degree equations, and systems of equations in one or more unknowns by the false position method and the direct resolution method. In this regard, it is interesting to note that Leonardo calls the direct method for the resolution of first degree equations "the method used by the Arabs, a very appreciable method since it allows us to solve many questions." This method, which is essentially taken from al-Khwarizmi's book on algebra, is nothing more than the algebraic method we use today to solve a given first-degree equation, although Leonardo once again presents it in a colloquial way and does not make use of symbolic expressions.

Among the innumerable problems contained in Chapter 12, we also find the solution to indeterminate equations such as the "bird problems" derived from Abu Kamil and other problems related to them. Finally, recreational problems are also presented, such as guessing a number that is in the mind of others, the checkerboard problem consisting of calculating powers of 2 up to 2^{64}, or finally, the well-known rabbit problem that gives rise to the famous Fibonacci succession of numbers.

The last three chapters again have a more theoretical character, although they are always enriched with numerous examples and practical problems. The presentation of demonstrations and the illustration of the execution of algorithms in an abstract style, supported by the use of literal symbols, become more and more present, and at the same time, the reference to the sources of Greek and Arabic mathematics is explicitly manifested.

Chapter 13 is entirely devoted to solving first-degree equations of the type $ax + b = c$ using the double false position method. By adopting two different values for the unknown x, two different values of c are obtained, and from the interpolation of these results, the correct value of the unknown[27] can be deduced. Leonardo calls this method for solving equa-

---

[27] Gino Arrighi, *Considerazioni sul Liber abaci di Leonardo Pisano*, in Morelli and Tangheroni (eds.), *op cit.*, p. 37.

tions *elcha:aym* (from the Arabic term *al-khata'ayn*, which means "two errors").[28] Al-Khwarizmi and the later mathematician al-Samawal al-Magribi (1130–1180), a Jewish convert to Islam who was born in Fes, Morocco, and lived in Baghdad, also wrote about the method of double false position in their works. The latter was the author of *Al-Bahir fi'l jabr* (the *Book of Enlightenment in Algebra*), in which he introduced the principle of mathematical induction and presented the powers of the binomial (and the triangle of Pascal) up to the power $n = 12$. Al-Magribi indicated that the method of the double false position was known to al-Biruni[29] (973–1048) as well as to the Indian mathematician Brahmagupta. In fact, it was also known in China, where it seems to have been discovered in around 100 CE and was called *ying pu tsu* ("too much and too little").[30]

Chapter 14 deals with topics that Leonardo himself claims are derived from Euclid's Book II and Book X: the distributiveness of multiplication, the square of a binomial, square roots, cubic roots, roots of roots, rational and irrational numbers, and so on.

Finally, Chapter 15 contains an exposition of properties and applications of proportions, taken from Euclid's Book II, the problems of plane and solid geometry (which Leonardo would take up and elaborate on in *Practica geometriae*), and most importantly a treatment of second-degree equations. This is substantially based on the algebra books by al-Khwarizmi[31] and his successors, such as Abu Kamil and al-Karaji, in relation not only to the six classes in which the equations are categorized but also to the resolution algorithms and finally to the rich set of exercises offered to the reader (as many as ninety-six). Typical problems of this kind are those (thirty-four) in which it is required to divide the number 10 into two parts that satisfy appropriate properties (for example, so that their product has

---

[28] Devlin, *op cit.*, p. 86.

[29] Al-Biruni, who was also born in Khorasmia, was an outstanding scientist: A mathematician, astronomer, astrologer, physician, and philosopher, he spoke several languages, including Sanskrit and Greek. Among his most important contributions is the calculation of the earth's radius (6,314.5 kilometers), an improvement on the value estimated by Eratosthenes.

[30] Devlin, *op cit.*, p. 103.

[31] Leonardo explicitly refers to al-Khwarizmi by noting "Maumeht" in the margin of the text.

a given value). One important new element is the fact that Leonardo also takes into account solutions of equations equal to zero and negative solutions and even identifies irrational solutions ($\sqrt{2}$, $\sqrt{7}$, etc.). Another interesting aspect is the terminology he adopts for unknowns and variables of algebraic equations. The square of the unknown is called *census* ("value", "sum of money"), a literal Latin translation of the Arabic word *mal* used by al-Khwarizmi,[32] while the unknown is called *radix* or *res* ("thing"),[33] a translation of the Arabic word *shay*; the constants (which al-Khwarizmi sometimes calls *dirhams*) are called *numerus*.

As we have stated, Leonardo explicitly indicates in some parts of the text which references he was inspired by, and we also have evidence that a significant percentage of the problems he proposed (which we will discuss in the next section) is derived from the works of al-Khwarizmi and other Arabic authors or from texts of the early medieval tradition. Unfortunately, we do not have more precise information on Leonardo's mathematical education.[34] In particular, we do not know to whom Leonardo is referring when speaking of his "wonderful education" in Bugia. Undoubtedly, Bugia, along with Tlemcen, was a cultural center of great importance in the 12th century and was home to important schools in various fields, from mathematics to philosophy. An interesting picture of the mathematical schools operating in the Maghreb in the 12th century has been painted by the historian Djamil Aïssani,[35] whose investigation shows that in the Maghreb at that time, several scholars were active with

---

[32] The translation of the term *mal* to the term *census* appears in the translation of al-Khwarizmi's algebra by Gherardo da Cremona.

[33] In Latin, Fibonacci uses the word *res*, which in Italian later translated as 'cosa'. This word lies at the origin of the fact that in the medieval and Renaissance centuries, mathematicians who devoted themselves to solving equations were sometimes called *cossisti*; see Ambrosetti, *op cit.*, p. 249. In fact, the German term *coss* derived from the Italian term *cosa*, and the expression *cossist algebra* was used to refer to a group of 'calculating masters' who gathered around the mathematician Christoph Rudolph (1499–1545).

[34] On the sources that inspired Leonardo Pisano's work, see also Jens Høyrup, *Leonardo Fibonacci and Abbaco Culture. A Proposal to Invert the Roles*, in "Revue d'histoire des mathématiques," 11, 2005, pp. 23–56.

[35] Djamil Aïssani, *Les mathématiques à Bougie médieval et Fibonacci*, in Morelli and Tangheroni (eds.), *op cit.*, See also Devlin, *op cit.*, pp. 65–79.

whom (or with whose students) Leonardo may have interacted. It has to be noted that they were operating in close contact with al-Andalus (the territory of Islamic Spain). In particular, it appears that in the 12th century, the mathematicians al-Qurashi (a Sevillian and an expert in the field of algebra), Abu Hassan (also an astronomer), and Abu Bakr (who worked in Ceuta, Marrakesh, and Spain and was the author of *Kitab al-kamil fi sinaat al-adad (Complete Book on the Art of Numbers)*), were active in Bugia. It seems that it is to al-Hassar that we owe the compound fractions Leonardo introduces in *Liber abaci* and the notation for simple fractions in which the numerator and denominator are separated by a horizontal line.

Liber Abaci: **The Daily Life of a Merchant**

Overall, the book contains a considerable number of exercises and problems accompanied by the presentation of the relevant solution algorithms, in several cases supported by demonstrations the author calls "Euclidean". There are about 440 problems of a practical nature from Chapter 8 onward (a conspicuous number of which are in Chapter 12), and there are also more than 150 exercises in numerical calculus, concentrated mainly in the first chapters and in Chapter 15: that is, in the chapters devoted to arithmetic operations and algebraic equations. When one examines these problems, it is not difficult to identify various authors from whom Leonardo drew inspiration, including, as noted earlier, Alcuin, al-Khwarizmi, and Abu Kamil.

One of the most interesting aspects of the applied problems is that — in addition to a review of the main mathematical techniques needed to deal with the problems themselves, of course — they offer a very comprehensive picture of the activities that merchants performed in that historical period. In his book, as discussed in the following, Leonardo addresses all the main daily activities a merchant had to perform, for which it was important to have skills in numerical computation, dealing with fractional numbers, calculating proportions, and solving first- and second-degree equations[36]:

---

[36] A discussion of the applied themes presented in *Liber abaci* can be found in Enrico Giusti, *Matematica e commercio nel Liber abaci*, in Id. (ed.), *Un ponte sul Mediterraneo. Leonardo Pisano, la scienza araba e la rinascita della matematica in Occidente*, Polistampa, Florence, 2002.

1. *Accounting management.* In one of the first chapters, Leonardo shows his interest in the mathematical training of merchants. Chapter 3, which is devoted to multiplication and addition, explains, for example, how a treasurer or ship's scribe is to keep accounts of a ship's expenditures and how numbers are to be recorded and put in columns in order to make correct sums. In particular, a sample table shows how expenditures in *lire*, *soldi*, and *denarii* are to be put in columns, and it is explained that the same method is to be used to deal with other currencies, "bezants, carats, Genoese *tareni*", or numbers relating to any other magnitude.[37]

2. *Buying and selling and bartering of goods.* Typical issues in this area include establishing the selling or buying price of a given quantity of goods given the price of another quantity that is perhaps expressed in different units. It is essentially a matter of calculating proportions by the method of three simple or three compound. The following are two problems that can be cited as examples (taken from chapters eight and nine):

100 pounds of pepper are worth *lire* $(11 + 9/20)$, what are 46 pounds + ounces $(5 + 1/4)$ worth?

Obviously, the solution is obtained (rule of three simple) by multiplying $(11 + 9/20) \times (46 + (5+1/4)/12)$[38] and dividing by 100. The result presented by Leonardo, in fractional form, corresponds to 5 *lire* + 6 *soldi* + $(4 + 1/2,400 + 1/96,000)$ *denarii*.

The following is proposed: 7 rolls[39] of pepper are worth 4 bezants[40] and 9 pounds of saffron are worth 11 bezants and you want to know how much saffron you can have for 23 rolls of pepper.

---

[37] See Tito Antoni, *Leonardo Pisano detto il Fibonacci e lo sviluppo della contabilità mercantile nel Duecento*, in Morelli and Tangheroni (eds.), *op cit.*, p. 46.

[38] One ounce corresponds to 1/12th of a pound.

[39] The roll was a Pisan unit of weight measurement; 100 rolls corresponded to about 50 kilograms.

[40] The bezant was the currency of Constantinople used in the Eastern Roman Empire, but it was also widespread throughout the Mediterranean area.

According to the method suggested by Leonardo, the following table is prepared:

Pounds of saffron	Bezant	Pepper rolls
?	4	7
9	11	23

Given the five known numbers, one must calculate the sixth (a rule Leonardo calls the "rule of six"). The problem is solved by "multiplying the two numbers at the ends of the row with three numbers and the number in the middle in the top row and dividing by the two other numbers." So, the result is pounds $23 \times 4 \times 9/7/11 =$ pounds $10 + 8/11 + 2/77$.

3. *Change of units of measurement of different countries.* As we know, the political fragmentation of Italy in the medieval period, particularly from the age of the communes, also resulted in an extreme diversification of the units of measurement used in the various cities. In many cases, the lack of homogeneity of measurement units, even between neighboring territories (such as Pisa and Florence) also arose from the fact that control of measurements was sometimes entrusted to private individuals who evidently had no interest in standardizing them. In addition, wide-ranging trade made it necessary for Italian merchants to also be familiar with the units of measurement used in North Africa (in Tunis as in Alexandria), the Eastern Roman Empire, Provence, and Spain. Leaving aside coins, which we will examine in the following, reference is made in *Liber abaci* to a great variety (several dozen) of units of measurement of weight, length, and capacity that were characteristic of different locations in Italy and elsewhere. As for units of weight, further diversification arose from the need to weigh bulky goods (for instance, in *cantari* or Pisan rolls, about 50 kilograms and about 0.5 kilograms), smaller goods (the pound was equal to 340 grams and a Pisan ounce was equal to 1/12th of a pound), or finally minute, valuable goods such as precious stones (whose weight was measured in *denari di cantare*, equal to 1/39th of an ounce; in carobs, equal to 1/6th of a

denaro; or in grains, equal to 1/4th of a carob).[41] The units of length were also differentiated: For example, the cane (used for land measurement) was equal to 5 arms (one arm corresponded to about 58 centimeters), while the Pisan mercantile cane was equal to 4 arms. *Liber abaci* also mentions the Genoese cane (equal to 9/10th of the Pisan one) and canes of Provence, Sicily, Barbaria, Syria, and Constantinople (all of the same length, equal to 8/10th of the Pisan cane). As units of length, Leonardo also refers to the palm (about 24 centimeters, corresponding to 1/10th of a cane), the piece (equal to 2 canes), and the *torcello* (equal to 6 canes). A typical example of the problems Leonardo posed in this area is as follows:

Again, if you want to turn 43 Messina rolls [of a product] into Pisan pounds you must first determine what proportion there is between Messina rolls and Pisan pounds. I believe the proportion is that one Messina roll corresponds to (2 + 1/4) Pisan pounds, so four Messina rolls correspond to 9 Pisan pounds. Then you set the problem as shown here and multiply the 9 × 43 (which is diagonally opposite it) and divide by 4; the quotient is 96 Pisan pounds and 9 ounces.

Pisan pounds	Messina rolls
9	4
?	43

4. *Exchange of currencies of different countries.* The coins mentioned by Leonardo are also extremely diverse. After the monetary reform implemented by Charlemagne in 794 — which imposed the system based on silver libra (later called *lira*) of 20 solids (later called *soldi*) and a soldo of 12 *denarii* throughout the Carolingian Empire — coins diversified as early as the 9th century. Besides they began to lose their effective value by being minted with alloys of gradually lower silver content. By the 12th century, twenty-eight Italian cities, including Pisa, Genoa, Bologna, Piacenza, Pavia, Lucca, Milan, and Venice,

---

[41] Sigler, *op cit.*, pp. 621–622.

were minting coins independently, although there was no shortage of attempts to establish agreements that facilitated exchanges.[42] For its part, in Sicily, during Arab rule, the minting of gold coins (the *tareno*, or *tarì*, corresponding to a quarter *dinar*) and small silver coins (called carobs)[43] had continued, and these coins also continued to be minted in Sicily during Norman rule. The coins that appear in the problems presented by Leonardo show the breadth of the commercial sphere to which he refers. In fact, Pisan, Bolognese, Venetian, Genoese, Paduan, Magalona (Occitania), and Barcelona money, regal money, imperial money, and so on appear in the problems, but so do coins of the Eastern Roman Empire such as the bezant, which was widespread throughout the Mediterranean area, and Islamic coins (the Garb bezant, the Saracen bezant, and the Massamutin, a gold bezant minted in Spain by the Almohads).[44] Again, we provide an example of a problem from *Liber abaci*:

An imperial *soldo* is worth 28 + 1/2 Pisan *denari*; one asks how many Bolognese *denarii* correspond to one Pisan *lira* knowing that one imperial *soldo* corresponds to 36 Bolognese *denarii*.

Here, too, the problem is solved by a simple proportion.

5. *Determining the alloys needed to mint coins of different values.* Minting coins with the correct silver and copper alloys was a problem that not only concerned city or state institutions but also directly affected the merchants themselves. In fact, in many cases, while moving from one city to another, merchants had to bring their own coins to the mint and have them minted again into local coins (with possible additions of silver or copper), a practice that seems to have been widespread even in

---

[42] Lucia Travaini, *Monete, mercanti e matematica*, Jouvence, Sesto San Giovanni, MI, 2003.

[43] It should be noted that in various cases units of weight and coins took on the same name as in the case of the pound (later to become the *lira*), the *tareno*, and the carob. This also happened in the English language, where the term pound continues to denote both weight and coin.

[44] The name *massamutin* derives from the *masmudah* tribe from which the Almohads came. Similarly, the coinage minted by the Almoravids took the name *marabotin* and spread to Spain and Portugal as *maravedi*.

antiquity. Mints were thus of primary importance in medieval cities, as is also shown by their location (in Florence next to the Palazzo Vecchio and in Siena even in the City Hall).[45] Mints could be managed directly by city (or state) authorities or contracted out to private individuals (merchants or bankers). It is therefore interesting that Leonardo devotes an entire chapter, Chapter 11, of *Liber abaci* to coinage alloys, considering various situations: alloys consisting of given amounts of silver and copper, alloys consisting of melted coins to which silver, copper, or both are added, and alloys consisting only of the melted-down given coins. An example of a problem related to the last case is as follows:

For example, a man has coins with 2 ounces [of silver] and coins with 9 ounces with which he wants to produce coins of 5 ounces. Write 2 and 9 on the same line and below between them write 5 as is shown.

Second coin	First coin
3	4
9	2
5	

The difference between 9 and 5 — i.e. 4 — you will write above 2, and the difference between 5 and 2 — i.e. 3 — you will write above 9. You will put in 4 parts of the lesser silver coins and 3 parts of the greater silver coins.

One interesting aspect of the chapter on alloys is in the last section of the chapter entitled *Rules for mixing similar things*, which is devoted to problems that are related in some way to metal casting. In this section, we find the well-known 'bird problems' introduced by Abu Kamil, which, as we will recall, correspond to solving indeterminate Diophantine equations. An example of the problems posed by Leonardo in this section is, in fact, a bird purchasing problem:

A man buys 30 birds, partridges, pigeons, and sparrows, for 30 *denari*. He buys a partridge for 3 *denari*, a pigeon for 2 *denari,* and a sparrow for 1/2 coin. One wants to know how many birds he buys of each species.

---

[45] Travaini, *op cit.*, p. 22.

A second example, in which, unlike in the bird problem, the solutions are not required to be integer, is the following:

A man buys 90 bushels of wheat, millet, beans, barley, and lentils in Constantinople for 21.25 bisants. Knowing that one hundred bushels of wheat sells for 29 bezants, barley for 25 bezants, millet for 22 bezants, beans for 18 bezants and lentils for 16 bezants, one wants to know how much he buys of each type of grain.

Leonardo turns the problem into a problem of alloys for coinage.[46] Slightly simplifying Leonardo's reasoning, we first relate the cost of grains to 90 bushels and obtain the respective costs 26.1, 22.5, 19.8, 16.2, and 14.4. Then, following his method of reducing the problem to a coin-casting problem, we can say that we have coins with 26.1 ounces (of silver) and with 22.5, 19.8, 16.2, and 14.4, and we want to make a coin with 21.25 ounces (of silver). To solve the problem, we average the costs (per 90 bushels) of the two most expensive types of grains and the three least expensive types of grains, obtaining 24.3 and 16.8 respectively. At this point, we can calculate how to mix the grains of the first type with the grains of the second type to obtain 90 bushels at a cost of 21.25 bezants. Solving a simple first-degree equation shows that we can use 59.3 percent of the first type (assuming we buy the same amount of the two types of grains) and 40.7 percent of the second type (again assuming we buy the same amount of the three types of grains).

6. *Division of investments and profits among partners in a company.* The formation of companies among merchants was an important chapter in medieval economic history, and so the information Leonardo gives us about this practice through the presentation of problems is particularly interesting. Companies of two, three, or four merchants are mentioned in a wide variety of contexts, as enterprises for the conduct

---

[46] Leonardo Pisano's ability to find a level of abstraction that allows him to assimilate two different problems like the grain problem and the alloy problem is further evidence of the Pisan mathematician's computational skill.

of commercial activities and the consequent division of profits, or for the purchase of goods or animals. In a problem posed to Leonardo by a wise man at the Mosque of Constantinople, the purchase of a ship by five partners is even hypothesized. A simple example of this type of problem is as follows:

It is proposed that two men form a company together in which the first man invests 18 *lire* and the other invests 25 *lire*. The company makes a profit of 7 *lire*. How much of the 7 *lire* will each of the two have?

The solution algorithm given by Leonardo (which is obvious to us) is laid out in the introduction to Chapter 10:

Write the number of shares of the first partner at the top of the table on the right; then on the same line in order toward the left put the shares of the other partners. Finally at the top left write the profit of the partner- ship. Then add up the shares of all the partners and keep the total. Then divide each of the partners' shares by the total and multiply by the com- pany's profit; thus you will have the share of the profit that each partner is entitled to.

Profit	Second partner	First partner	
7 lire	25 lire	18 lire	Investment
	25/43	18/43	Share of profit
	4 + 3/43	2 + 40/43	Profit

The profit shares due to the two partners are therefore *lire* 2 + 40/43 and *lire* 4 + 3/43.

In addition to these types of problems, the book offers many other insights that we will not dwell on here: problems related to a journey dur- ing which earnings are made and expenses incurred, problems of the consumption of food, problems of working on a given number of days and compensation, paying the costs of transporting goods on a ship, borrowing money and paying interest, and so on. Of particular note in terms of the complexity of the required mathematics are the commercial arithmetic problems concerning the calculation of interest (credit or debit), which require the solution of exponential equations, which Leonardo solves brilliantly, even though he does not know the concept of the logarithm. Overall, this is an extremely wide-ranging review of activities that offers Leonardo the opportunity to show not only his mathematical skill but also his knowledge of the Mediterranean mercantile environment. As we examine these aspects, in fact, we cannot forget that we are talking about Pisan merchants — that is, merchants of what was one of the main trading powers of the Mediterranean until the clash with rival Genoa and the consequent defeat reduced its ambitions at the end of the 13th century. The range of goods, coins, units of measurement, and places mentioned makes us realize that the mercantile space to which Leonardo refers is the entire Mediterranean area. There is mention of selling clothes and buying cotton in Barbaria, exchanging goatskins for cowskins in Bugia and Ceuta, selling precious stones and pearls in Constantinople, selling linen in Syria, and buying pepper in Alexandria.

To the problems dealing with these practical aspects of merchant life, Leonardo adds a wide range of problems that are less realistic but belong to the mathematical tradition. Among them, we encounter problems already present in Alcuin's text or that had even been known since antiquity, such as sums of arithmetic progressions and nursery rhymes that require the sum of geometric progressions. In addition, one finds problems of a geometric type (calculation of surfaces, volumes, etc.) that Leonardo later deals with in greater depth in the *Practica geometriae*; problems involving filling and emptying vessels; "tree problems",[47] in which one has to calculate the height of a tree knowing that a certain part is underground; problems involving proportions, powers, square roots, and perfect numbers; problems requiring the solution of second-degree equations and so on; and finally problems that require guessing a number after various transformations have been performed on it. One of the frequently proposed recreational problems is the "purse problem",

---

[47] Problems concerning the height of trees are similar to those concerning the age of a man (see the riddle on Diophantus' tomb) and fall into the category of the so-called 'heap problems': that is, problems requiring the solution to first-degree equations.

which requires the calculation of the amounts of money contained in a purse and those possessed by some men by knowing the existing ratios among them.[48]

Among the problems proposed by Fibonacci that do not correspond to the computational needs of the merchants but to recreational mathematical curiosities, we encounter the famous "rabbit problem":

> A man puts a pair of rabbits in a place surrounded by walls to find out how many pairs of rabbits would be born in a year because by their nature in each month from a pair another pair is born and they start breeding from the second month.

The solution to the problem[49] corresponds to the well-known Fibonacci succession — $F_0 = 1$, $F_1 = 1$, $F_{n+2} = F_n + F_{n+1} (n \geq 0)$ – which, as many know, recurs in various areas of nature from botany to biology and finds applications in multiple human activities, architecture, art, and music. As an example, let us recall that the ratio of one Fibonacci number to the next converges to the (irrational) value of the golden ratio $f = 0.618033$, which corresponds to the inverse of the ratio of the length of a segment to the length of its golden section.[50] The structure of Greek temples was based on this ratio, and it has inspired artists and musicians in their compositions in recent times.

Indo-Arabic Computational Methods and 'Algorisms'

As mentioned previously, Leonardo Pisano's work marked a fundamental turning point in the landscape of medieval calculus and gave impetus to

---

[48] The purse problem recurs eighteen times in *Liber abaci*. It also appears in *Ganita sara sangraha* written by the Jain mathematician Mahaviracarya (800–870). See Devlin, *op cit.*, p. 92.

[49] A similar problem involving the reproduction of cows and calves was posed later, in the 14th century, by the Indian mathematician Pandit Narayana. In his case, the succession obtained is as follows: $N_0 = 1$, $N_1 = 1$, $N_2 = 1$, $N_{n+3} = N_n + N_{n+2}$ ($n \geq 0$). See Alfred S. Posamentier and Ingmar Lehmann, *I (favolosi) numeri di Fibonacci*, Muzzio, Rome, 2010.

[50] Given a segment of length l, the golden section is that part of the segment of length x such that $l : x = x : (l - x)$.

the mathematical studies that were later developed in the following centuries.[51] Despite Fibonacci's undisputed role, however, we must make it clear that his was not the only contribution to bring about the spread of Indo-Arabic methods of calculation into everyday life in Europe. In the first half of the 13th century, other scholars played a role in popularizing arithmetic calculus and algebra, compiling works that although less ponderous than *Liber abaci* were widely circulated, adopted, and used in academia for a long time.

The term that unites these texts is *algorism*. As we have already had occasion to say, it comes from a deformation of al-Khwarizmi's name, but this origin, which identifies its Islamic source, was forgotten fairly early. As early as the 13th century, in fact, various authors attributed a false etymology to the term according to which it derived from the name of a king or philosopher (*Algus*) and the Greek word *arythmos* ("number").[60] It is curious that in 1307, Jacopo da Firenze, while acknowledging the Arabic origin of the term when writing his *Tractatus Algorismi*, provided another etymology: "And know that we call it *algorismus* because this science was chiefly made in Arabia, and those who found it were similarly Arabic. And the art is called *algo* in the Arabic language and the number is called *rismus* and therefore it is called *algorismus*".[52]

Among the earliest authors who produced mathematical works of particular interest in the 13th century was the Frenchman Alexandre de Villedieu (Alexander de Villa Dei, 1175–1240), a religious man (perhaps

---

[51] Until the sixteenth century, "the *Liber abaci* was held in such high regard among mathematicians that it was often cited by authors to ennoble their own writings. Even in 1494 Luca Pacioli cited it among his sources"; see Gavagna, *Leonardo Fibonacci*, in Clericuzio and Ricci (eds.) *op cit.*

[52] Ambrosetti, *op cit.*, p. 233. The first note to the 1841 edition of *Carmen de Algorismo* reads, "*Haec praesens ars dicitur algorismus ab Algore rege ejus inventor, vel dicitur ab algos quod est ars, et rodos quod est numerus; quae est ars numerorum vel numerandi, ad quam artem sciendum inveniebantur apud Indos bis quinque (id est decem) figurae*" (This art is presently called "algorismus" by its inventor, king Algore, or by the word "algos" which means art and "rodos" which means numerus; this is the art of numbers, to learn which the Indians have found twice five (that is ten) figures); see. Alexander de Villa Dei, *Carmen de Algorismo*, in James Orchard Halliwell (ed.), *Rara Mathematica, or a Collection of Treatises on the Mathematics subjects connected with them, from ancient inedited manuscripts*, London, 1841, p. 73.

a Benedictine monk) who taught in Paris and Brittany. In around 1202, Villedieu published a rather curious mathematical text that was widely circulated at the time entitled *Carmen de Algorismo*. According to some historians of mathematics, *Carmen* probably did more to make Indo-Arabic numerals known than any other work of the time. It is a didactic poem in Latin hexameter devoted to Indo-Arabic numerals and arithmetic operations. The poem begins with the words *Haec algorismus ars praesens dicitur; in qua talibus Indorum fruimur bis quinque figuris 0, 9, 8, 7, 6, 5, 4, 3, 2, 1* (This art is presently called "algorismus", in which we make use of twice five figures ...). Note that unlike the presentation of the figures provided by Leonardo Pisano, who introduces the nine Indian digits separately from the 0, ten digits including the 0 are presented in *Carmen*. After the presentation of the figures and an illustration of the positional system, the 'seven' operations are introduced: *Addere, subtrahere, duplareque dimidiare/ Sextaque dividere est, sed quinta est multiplicare/ Radicem extrahere pars septima dicitur esse* (that is, addition, subtraction, doubling and halving, division, multiplication, and square root extraction). Arithmetic progressions are also presented in some manuscripts. The relevant algorithms are described for all the operations, but it should be kept in mind that these algorithms refer to operations performed on a sand tablet, which requires erasing and rewriting the digits of one of the two operands and the intermediate results.[53]

Another text that effectively contributed to the dissemination of Indo-Arabic numerals, thanks to the fact that it was introduced and used for a long time in academia, was the work entitled *Algorismus vulgaris* by John of Halifax (ca. 1195–1250), who was also called John of Holywood, after the name of the monastery where he was a canon of the order of St. Augustine, but was generally known as Johannis de Sacrobosco. What little is known of his life indicates that he was probably born in England and that he studied first at Oxford and then, from about 1220, in Paris, where he later taught mathematics and astronomy. His most challenging work, which was published in 1230, was *Sphaera*

---

[53] Nadia Ambrosetti, Algorithmic in the 12th Century: The Carmen de Algorismo by Alexander de Villa Dei, *3rd International Conference on History and Philosophy of Computing (HaPoC)*, Pisa, October 2015, pp. 71–86 (HAL-01615308).

mundi, the fundamental reference for European astronomy for about four centuries, which described the Earth and its place in the universe according to the Ptolemaic system. Another important work of his, published in 1235, was *Libellus de anni ratione seu computus ecclesiasticus*, which was devoted to calendars and the study of time, for the measurement of which Sacrobosco introduced the time subdivision in seconds, minutes, and hours. His work *Algorismus vulgaris*,[54] also known as *Tractatus de arte numerandi*, which he wrote in 1240, was devoted to arithmetic calculus with Indo-Arabic numerals. It was given great emphasis and was adopted as a text in university circles for about three centuries. It should be noted that in his presentation of the digits, Sacrobosco distinguishes the digits from 9 to 1 with respect to 0, which he calls *theta, vel circulus, vel cifra, vel figura nihili quia nihil significat sed locum tenens* and for which the fundamental role of "placeholder" in the positional system is recognized. It is also significant that Sacrobosco points out that when we write numerals, we use the left-handed verse of Arabic, while in reading we use the usual right-handed verse. Although it is short, and still based on a rhetorical exposition of mathematical concepts, Sacrobosco's work is quite comprehensive as, in addition to the four arithmetic operations, it includes the calculation of progressions and the extraction of square roots and cubic roots.

The last thirteenth-century author we believe it is important to cite for his contributions to mathematics, and in particular to arithmetic and algebraic calculus, is Jordanus de Nemore (or Jordanus Nemorario). Very little biographical information is known about him. Some authors consider him to be Italian; others identify him with Jordanus of Saxony, the first successor of St. Dominic as general of the Dominican order from 1222 to 1237. Still others believe that the appellation "de Nemore" indicates that he worked in the city of Nemours; finally, others attribute a teaching position at the University of Toulouse to him. In any case, four of his works appear in a list of books that were of interest to the Cathedral of Amiens (the *Biblionomia* of Richard de Fournival). These books were compiled before

---

[54] Sacrobosco also gives a false etymology to the term *algorism*, suggesting that the word derives from the name "Algus" of a philosopher; see Joannis de Sacro-Bosco, *Tractatus de arte numerandi*, Halliwell (ed.), *op cit.*, p. 1.

1260, and thus reveal that he played an important role in the European culture of the time and carried out wide-ranging scholarly activity. Historians point to him as "second only to Fibonacci in medieval mathematics."[55]

In fact, Jordanus Nemorario can rightly be defined as a mathematical physicist; in fact, his most interesting works, which are outside the scope of the topics covered in this book, are on mechanics and, in particular, weights and levers (*Elementa super demonstrationem ponderum, De ratione ponderis,* etc.). According to historians, these works, which refer to the studies of Archimedes and Thabit ibn Qurra, laid the foundation for studies of statics in the Middle Ages. For example, a fundamental result in them deals with the law of the sloped plane that determines force as a function of the slope grade.

Jordanus' works had a profound impact in the more strictly mathematical sphere and appear to have been reproduced and printed as late as the 16th century. In this field, in addition to texts on geometry and stereographic projection, we can identify three main strands. The first consists of treatises on "algorisms": *Demonstratio de algorismo* and *Demonstratio de minutiis* (which appeared in different versions) devoted to the presentation of Indo-Arabic methods of calculation, with reference to integers and fractions, respectively. The second consists of the arithmetical work *De elementis arismetice artis* (which was also published in various versions), made up of ten books in which four hundred propositions based on the works of Euclid and Boethius are expounded, with the ambitious goal of producing an all-encompassing reference text comparable to Euclid's *Elements*. The most interesting aspect of this work is the presence of actual demonstrations accompanying the propositions. One example concerns the product of a binomial:

> If a number is divided into two parts the product of the whole by itself
> is equal to the product of each part by itself and twice the product of one

---

[55] Edith D. Sylla, Book Review: De numeris datis. Jordanus de Nemore, Barnabas Bernard Hughes, in *Isis*, 74, 1, 1983. See also Jens Høyrup, Jordanus de Nemore, 13th Century Mathematical Innovator: An Essay on Intellectual Context, Achievement, and Failure, in *Archive for History of Exact Sciences*, 38, 1988, p. 308.

part by the other. [The proof in symbolic terms follows]. Let the number *ab* be divided into *a* and *b*. I say that the product of *ab* by itself is equal to the product of *a* by itself and *b* by itself and the double of *a* by *b*.

The third and final strand concerns algebra — indeed, we might say advanced aspects of algebra. The text *De numeris datis*, which according to historians anticipates the first works on algebra that appeared in the 16th century by three hundred years, is highly innovative both in terms of the presentation of problems and the methods of solution, although it is based on the contributions of Arabic algebra texts. In it, the abstract formal treatment (although not yet in fully symbolic form) of equations is constantly accompanied by numerical examples.

The treatise, which was supposedly written in 1225, consists of four books and, according to the historian Nadia Ambrosetti, "is characterized by a strong Euclidean heritage."[56] In addition to this reference, the text naturally draws on the legacy of Arabic treatises on algebra. Many problems and examples proposed by Jordanus are taken from the works of al-Khwarizmi, Abu Kamil, and al-Farabi. The classification of second-degree equations is also inspired by al-Khwarizmi's text. Other insights appear to be drawn from *Liber abaci*, although it is unclear whether Jordanus had the time to study Leonardo Pisano's work, which had been published only a few years earlier. In the work *De numeris datis* he presents problems in their generality, using letters instead of numbers, drawing inspiration from Euclid, who in books VII–IX of *Elements* represents numbers with linear segments. By interpreting the symbols *a* and *b* as numbers and at the same time as segments, Jordanus can represent the sum as a concatenation of two segments and thus write *ab* meaning the sum of the two numbers *a* and *b*. This partially symbolic style allows the adoption of a deductive analytical approach consisting of the following steps: (i) formal statement of the problem and construction of an equation; (ii) transformation of the equation in order to place it in a canonical form; (iii) calculation of the solution, with reference to a numerical example. This approach is the essential merit of Jordanus' works and, as noted earlier, is similar to that which would later be introduced in the 16th century.

---

[56] Ambrosetti, *op. cit.*, p. 238.

As an example, let us consider the following problem[57]:

If you know the ratio of two numbers together with the sum of their squares then you can know both numbers. Let *a be* the ratio of *x* and *y*. Let *d be* the square of *x* and *c* the square of *y*, and let *b* be the sum of *d* and *c*. Now the ratio of *d* and *c* is the square of the ratio of *x* and *y*. The ratio of *x* and *y* is known so the ratio of *d* and *c* is known. Add 1 and call it *e*. The equation to solve [in canonical form] is then *e* for the square of *y* equals *b*.

In modern terms, we would write the following:

$$x/y = a$$
$$x^2 = d; \ y^2 = c;$$
$$d + c = b$$
$$d/c = a^2$$
$$d/c + 1 = e$$

The equation to solve is $ey^2 = b$ from which we derive $y = \sqrt{[b/(a^2 + 1)]}$. The symbolic presentation of the problem is followed by a practical numerical application:

For example let 2 be the ratio of two numbers and the sum of their squares be 500. Now since the square of one number is 4 times the square of the other it follows that 500 is 5 times the square of the other number and so the other number is 100. The root of 100 is 10 and this is the smallest number and the largest is 20.

Another problem that can be solved with a first-grade equation is the following [58]:

If the quotient of two parts of a number is known, then the parts can be found. Let *c* be the quotient of *a* and *b*. Increase it by 1, you get *d*. Since

---

[57] Jordanus de Nemore, *De numeris datis. A critical edition and translation by Barnabas Hughes*, University of California Press, Berkeley, CA, 1981, pp. 5–6.
[58] *Ibid*, p. 18.

the product of b and c is a then the product of b and d is ab [i.e. $a + b$], so divide ab by d and you get b.

In modern terms, we would write the following:

$$a + b = g$$

$$a/b = c$$

$$c + 1 = (a + b)/b = d$$

$$b\,d = a + b = g$$

$$b = g/(c + 1)$$

Again, the symbolic presentation of the problem is followed by a numerical example:

For example, let 4 be the quotient of the two parts of 10. Increase by 1, you get 5. Divide 10 by 5 and you will get 2, one of the parts [being the other 8].

Chapter 6

Abacus Schools and Mercantile Life in the Late Middle Ages

The Abacus Schools

Leonardo Pisano's *Liber abaci* was the first mathematics text to explicitly address mercantile life, and it revolutionized the computational methods used by the merchants. After Fibonacci, being a merchant and mastering the abacus in the sense of computational skill were considered to be intimately linked to the point that in the anonymous treatise *Larte de labbacho* (a manual of about sixty pages published in print in 1478, also known as *Aritmetica di Treviso*), we read, "l'arte della merchadantia chiamata vulgarmente l'arte de l'abbacho" (the art of the merchants, commonly known as the abacus art).[1] However, despite the important role of *Liber abaci* and the other coeval works we have discussed — such as Villedieu's and Sacrobosco's books of "algorisms" — in the spread of Indo-Arabic methods of calculation in the mercantile circles of Europe, the establishment of these methods still stalled for a long time before they were fully adopted, because of both a natural resistance to change and mistrust and fear of fraud. It was not until the 14th century that the "revolution" in Arabic mathematics began to take hold for good under the impetus of demands from the mercantile sphere. This was contributed to in a decisive

---

[1] See Ugo Tucci, *Manuali d'aritmetica e mentalità mercantile tra Medioevo e Rinascimento*, in Morelli and Tangheroni (eds.), *Leonardo Fibonacci*, p. 53; see also Ambrosetti, *L'eredità arabo-islamica*, p. 259.

way by the abacus schools that were established in Italy (first and foremost in Tuscany) and other countries, attendance at which became an essential step in the careers of merchants.

As we have seen with regard to the city of Pisa, from the 12th century onward, trade between the different realities facing the Mediterranean (Italy, Provence, Spain, the Byzantine Empire, the Latin kingdoms in the Middle East, and the Islamic countries on the southern coast) developed considerably. Not surprisingly, this period has been the subject of intense study by historians. "The economic rebirth of Europe coincided with the commercial revolution of the 12th century": These are the opening words of Armando Sapori's volume *La mercatura medieval,*[2] in which he employs a rich documentary apparatus to examine the sustained growth of mercantile activity in Western Europe in the 12th to 14th centuries and sketches the figure of the merchant (Italian, in particular), his training, his culture, and his way of thinking and operating.

The merchant's trade involved intensive calculation, which gradually became more and more demanding: keeping accounting books (which in some cases also assumed a legal value), updating the ledgers in which the accounts pertaining to each correspondent (customer, supplier, etc.) were maintained on a daily basis, managing shares in a company, and so on. Of course, large commercial enterprises, especially large banks,[3] also had a division of labor among employees, so some clerks who were able to perform computations were dedicated exclusively to keeping the accounts, although mathematical skills were a necessity for all merchants. Being good at the art of the abacus turned out to be one of the requirements for a merchant who wanted to be successful in his work.

This propensity for mathematics on the part of Italian merchants is emphasized by the historian Armando Sapori: "This Italian merchant, of middle age, whom we saw as a constant and faithful annotator of all the facts of his life and of the news of interest that came to his attention, wanted to be most exact in his calculations and clear in his accounting, out

---

[2] Armando Sapori, *La mercatura medievale*, Sansoni, Firenze, 1972.
[3] In the first half of the 14th century, the Bardi bank was one of the most powerful in Europe, with branches in eleven cities in Italy and in major European cities (including Avignon, Barcelona, Bruges, Cyprus, Marseilles, Paris, and Seville).

of the conviction that exactness and clarity were indispensable in the conduct of mercantile affairs, and he implemented this by possessing an
adequate mathematical culture."[4]

A fundamental role in mathematical education was played by the
abacus schools. In the 13th and early 14th centuries, "there was a transi-
tion from the convent school to the communal school [...] which proceeded in parallel with private schools."[5] Along with the grammar schools
in which Latin, letters, and rhetoric were taught, the abacus schools corresponded to an intermediate level of study that followed a more elementary stage, represented by learning how to read and write.

On the basis of documentation from the first half of the 14th century —
interesting information that can be found in the "record books" and
"account books" of Florentine families — it appears that boys (unfortunately only boys) went to abacus school at the age of eleven and stayed
there for more than two years.[6] Another accurate, albeit later, account is
provided by Bernardo Machiavelli, Niccolò's father, who wrote,
"I remember how this day, the 3rd of January 1479, I delivered my son
Niccolò to master Pier Maria, who will teach him the abacus." At that
time, the son was ten years and eight months old, and he completed his
abacus studies in one year and ten months.[7] The percentage of young men
who attended abacus schools was very significant, especially in major
mercantile cities such as Florence and Venice. According to a passage in
Florentine Giovanni Villani's oft-quoted *Cronica*, in 1338, "We find that
there are between eight and ten thousand boys and girls learning reading.
Between one thousand and one thousand two hundred boys are learning
abacus and algorismus in six schools. And those who stand to learn grammar and logic in four large schools, between five hundred and fifty and six

---

[4] See Armando Sapori, La cultura del mercante medievale italiano, in Gabriella Airaldi
(ed.), *Gli orizzonti aperti. Profili del mercante medievale*, Scriptorium, Turin, p. 148.
[5] See Sapori, *op cit.*, p. 50.
[6] See Armando Sapori, *Studi di storia economica. Secoli XIII-XIV-XV*, vol. I, Sansoni,
Florence, 1982, pp. 67–68, note 1.
[7] See Elisabetta Ulivi, Scuole e maestri d'abaco in Italia tra Medioevo e Rinascimento, in
Enrico Giusti and Raffaella Petti (eds.), *Un ponte sul Mediterraneo. Leonardo Pisano, la
scienza araba e la rinascita della matematica in Occidente*, Polistampa, Florence, 2002,
pp. 121–159.

hundred." Taking into account that there were six abacus schools operating in Florence at that time, we can say that each of them had an average of one hundred and seventy to two hundred students[8] and, based on Villani's assessments, it is not difficult to calculate that the percentage of males attending abacus schools made up about 25 per cent of the youth of the time. The role of abacus schools did not end with the education of younger boys, however: The most influential schools, such as the Rialto School in Venice where Luca Pacioli, a well-known 15th-century mathematician, studied, were preparatory to university.

In the 14th century, the number of abacus schools (or workshops) that were active in Italy was very significant. We can count abacus schools in Tuscany and in the main Italian cities (Genoa, Modena, Brescia, Siena, Lucca, Pistoia, Pisa, Arezzo, Florence, etc.). Public schools — for which, as we will see later, the city authorities recruited and paid the teachers — were more widespread, but in cities with more mercantile activity, there was no shortage of private institutions. A closer look at Florence's schools tells us that in the 14th and 15th centuries, eight abacus schools sprang up in the Santa Maria Novella district, three of them in the Santa Trinità area. One particularly important school was the so-called Bottega di Santa Trinita, which was located in front of the church of the same name and was where Paolo dell'Abbaco, who was, as we shall see in the following, one of the best-known masters of the time, taught. Some of the most important abacus masters taught in the same school. Six other schools were opened in the Santa Croce district — three in the San Giovanni quarter and three in the Santo Spirito quarter. The protagonists of this educational system were, of course, the teachers — the abacus masters. Their names are little known to most people because for centuries, especially since the 17th century, they were not considered worthy of being remembered. It is only recently, thanks in part to a renewed interest among specialists in the computational aspects of mathematics, that studies by historians, including in particular Gino Arrighi, Raffaella Franci, Laura Toti Rigatelli, and Elisabetta Ulivi, have begun to return this important aspect of medieval life back to its proper prominence.

---

[8] *Ibid.*

Florentines played an important role among the abacus masters. Florence's abacus masters often also taught in other municipalities and they seem to have been highly regarded. Indeed, in a document of the town of Arezzo in 1451, we read the following: "Given that knowledge of arithmetic is very useful and necessary in every Republic, as the experience of our magnificent Florentine lords teaches, who have surpassed all others with this doctrine, [...] it is a good and necessary thing that the Aretine youth should learn this science."[9]

Some seventy abacus masters are believed to have operated in Florence over two centuries. One of the earliest documents referring to an abacus master dates from 1283. It mentions Jacopo dell'Abaco, perhaps the author of *Tractatus algorismi* (1307), one of the oldest abacus treatises, which will be discussed later. Also very active in the late 13th and early 14th centuries were the brothers Ranieri (or Neri) di Chiaro and Gherardo di Chiaro, who also taught in Siena. The aforementioned Paolo dell'Abbaco had other important masters among his students, such as Antonio de' Mazzinghi, who inherited his books and astrological instruments, and whose importance is linked to *Trattato di Fioretti*, an algebra text he published in 1373. Biagio di Giovanni, a disciple of Antonio de' Mazzinghi was another important later master.

The profession of abacus master often extended to several members of a family. For example, one family of abacists who operated in the 14th century was that of Maestro Moro, who was followed by his son Berto, his grandson Francesco, and his great-grandson Bartolomeo. Another example is the Corbizzi family: The first master in the family was Maestro Davizzo, who was active in the early 14th century and was followed by his sons Tommaso (who also taught in Siena) and Giovanni (the author of an abacus book in 1339), and his grandsons Bernardo and Cristofano. The largest and most influential family of Florentine masters began with Master Luca di Matteo.[10] Information on this comes to us from the land

---

[9] *Ibid.*

[10] Pier Maria Calandri, *Tractato d'abbacho*, edited and with an introduction by Gino Arrighi, Domus Galilaeana, Pisa, 1974, p. 15. See also Elisabetta Ulivi, *Gli abacisti fiorentini delle famiglie del Maestro Luca, Calandri e Micceri e le loro scuole d'abaco*, Olschki, Florence, 2013.

registry of Florence, which is a valuable source of information on families and trades. A disciple of Biagio di Giovanni, Luca wrote the fine *Arte d'abacho*. His son, the abacus master Giovanni, was the author of *Libro sopra arismetricha*. Master Calandro was a son of one of Luca's daughters, and they were followed by the masters Filippo Maria Calandri (author of *De arimetricha opusculum*, which was published in print in 1492 and dedicated to Giuliano de' Medici) and Pier Maria Calandri (author of *Tractato d'abbacho*).

Abacus masters, like merchants, also used to form partnerships to run schools together and divide the proceeds. Notarial archives record these kinds of contracts. Typically, the younger ones served an apprenticeship period as "assistants" and later entered into partnerships with the older masters or went on to teach at other schools. Some contracts made by abacus masters are a useful tool for understanding what their teaching activities consisted of, how the activities were divided chronologically between different subject areas, and how pay accompanied the various stages of teaching activities.

One interesting example that offers insight into what subjects were taught is the curriculum followed in the school of Cristofano di Gherardo di Dino. It is presented in his *Libbro d'anbaco* in 1442 and is called *al modo di Pisa*[11] (in the way of Pisa). First of all, the students were instructed to "make the figures i.e. 9, 8, 7, 6, 5, 4, 3, 2, 1". Then, there was digital calculus (*lo ponere alle mano*: that is, to put the units and tens in the left hand and the hundreds and thousands in the right). Then came the four operations on integers, followed by multiplication and addition of fractions ("broken" numbers). Then, it was on to application topics: the calculation of interest (*meritare a capo d'anno*),[12] then the exchange of coins and bartering (*le mute*), then geometric calculations (*misurare le terre*), and then calculations related to coins and alloys (*aconsolare et alleghare delli arienti*).

Other subjects, such as the rules of the simple three and the compound three and the proof of nine, appear in similar lists. Apart from the notions

---

[11] Ulivi, *op cit.*

[12] The expression means "to earn at the end of the year": that is, to compute interest using the system of capitalization at the end of the year; see Sapori, *op cit.*, p. 148.

of elementary geometry that were taught in schools, the heart of mathematical teaching was based on the simple notions of arithmetic and algebraic calculus, which we now identify with the solution of first- and second-degree equations with integer or fractional coefficients and which thus substantially corresponds to the mathematics of today's junior high school.

Paolo dell'Abbaco

The mathematics taught in abacus schools and abacus books was essentially computational. Even the geometry in abacus books is all computationally oriented, unlike Euclidean geometry, which was instead aimed at demonstrating the properties of the mathematical entities being studied. This computational approach strongly marked the mercantile culture of that period, and despite its limitations, led to the final establishment of Indo-Arabic methods of calculus. The medium used for teaching calculus in abacus schools was abacus books. In the period we are talking about, the 14th and 15th centuries, the production of this type of text developed remarkably. About two hundred and twenty abacus treatises and texts on elementary geometry produced in Italy between 1276 and 1500[13] are known, written in vernacular with different regional characteristics, many of them published in Tuscany. Ten were edited by the historian of mathematics Gino Arrighi. Generally, these texts were written by abacus masters and inspired by *Liber abaci* and *Practica geometriae* of Leonardo Pisano, but with a more concise and elementary presentation of the subject. The books were mainly written for students, covered both the fundamental topics of arithmetic (a description of Indo-Arabic digits and sometimes digital calculus, algorithms for the four operations on integers and on fractions, rule of simple three and compound three, and verification methods such as the proof of nine) and some typical applications of mercantile interest (transformations of units of length, weight, capacity, etc.; barters; the calculation of both simple and compound interest; the calculation of the losses and profits of partners in a company; the exchange

---

[13] Høyrup, *Leonardo Fibonacci and* Abbaco *Culture*, p. 35.

of coins; and the calculation of percentages of metals in alloys), and thus could also be useful to those who as tradesmen, merchants, land survey-ors, or municipal officials wanted rapid solutions to the practical problems at hand. Sometimes, abacus books contained recreational problems in addition to geometric, algebraic, and number theory problems. There is evidence of the spread of this kind of work in southern Europe as early as the early 14th century, and it is of interest to mention a few examples of the abacus books and treatises on arithmetic published in France. Two such texts, both published in Montpellier, Languedoc, are *Tractatus algorismi*,[14] which was written in 1307 by Jacopo da Firenze, and *Liber habaci*,[15] which was written (or transcribed) in 1310 by Paolo Gherardi (or Gerardi). These works — as well as the later *Trattato di tutta l'arte dell'abacho* which was published in Avignon around 1334, but the author of which is uncertain — not only highlight the contact existing between Tuscany and France but also seem to attest to the existence of an "abacus culture" outside Italy. In particular this culture was present in the Catalan-Provençal area, and thus, perhaps, not ascribable to the legacy of Leonardo Pisano, or even as the historian Jens Høyrup would have it, a culture that may itself have influenced Fibonacci's writing.

With regard to the work of Jacopo da Firenze (an author about whose identity there is a great deal of uncertainty, despite the research conducted by Elisabetta Ulivi), the content is entirely consistent with the classic typology of abacus books. The first chapter, in which the Indo-Arabic digits, zero, and the positional representation of integer numbers are pre-sented, is followed by seven chapters devoted to the four operations with tables of products and squares, even of very large numbers, and two fur-ther chapters devoted to operations between fractions. Next (Chapters 11–13), we have problems devoted to proportions (rule of three) and their applications in calculations related to currency exchange, simple interest, and the value of different quantities of commodities (typically pepper and grain). Chapter 14 (which resembles Chapter 12 of Leonardo Pisano's

---

[14] Jacopo da Firenze, *Tractatus algorismi*, edited by Jens Høyrup, Roskilde University, Roskilde, 2007.

[15] Paolo Gherardi, *Opera matematica. Libro di ragioni. Liber habaci*, edited by Gino Arrighi, MPF, Lucca, 1987.

Liber abaci) contains a variety of classical problems (purse problems, problems of galleys traveling in different directions at different speeds, the calculation of the height of trees of which a part is buried, will problems, etc.) and recreational problems. Finally, after five chapters devoted to plane and solid geometry, we have two that include problems regarding metal alloys and coins, with a list of one hundred and thirteen coins (Italian coins, but also bezants, marabotins, and massamutins as well as Aragonese, French, English, and German coins).

The twenty-five chapters of *Liber habaci* by Paolo Gherardi (a Florentine merchant and abacus master who lived between Florence and Montpellier) also follow a classic pattern consisting of a series of initial chapters devoted to operations between whole and fractional ("broken") numbers and a series of chapters of an applied nature devoted to calculating interest, bartering, exchanging coins, and calculating alloys. In addition, some chapters are devoted to geometric problems and, unusually, there are some concluding chapters that deal with elementary concepts of astronomy and hints of history.

More interesting than *Liber habaci* is another text Gherardi published in 1327, *Book of Reasons*. Unlike the previous text, only Indo-Arabic numerals are used in this work instead of Roman numerals. *Book of Reasons* also follows the classic pattern in the presentation of the material, but additionally includes a series of about twenty algebraic problems on the six canonical cases of second-degree equations to which, however, he also adds eight cases of third-degree equations. The problems are preceded by the solution rules and followed by a series of numerical case studies. An example of a rule referring to equations of a particular structure is as follows:

> Rule of things. When the *censi* and the things are equal to a number, we should divide by two the *censi* and then divide the things by two and then such halving we should multiply by itself and the thing is worth the root of that sum minus the halving of the things.

Remembering that the *thing* is the unknown and the *census* is its square, the rule states that if you have an equation of the type $x^2 + bx = c$, the value x is given by the expression $x = \sqrt{[(b/2)^2 + c]} - b/2$.

Among the abacus books published in the first half of the 14th century, Paolo dell'Abbaco's *Trattato d'aritmetica* plays a particularly important role. Paolo Dagomari,[16] called Paolo dell'Abbaco — but also called Geometra, Arismetra, and Strolago — was born in Prato in 1281. He was a pupil of another abacus master, Blaise "the Elder", who died in 1340 and was himself the author of *Tratato di practicha d'Arismetricha*.[17] For a long time, Paolo served as a teacher in Florence at the Santa Trinita abacus school. A manuscript of the time attributes to him "ten thousand excellent pupils." His school is also said to have been attended by Dante Alighieri's son Jacopo,[18] with whom Paolo exchanged some sonnets. After his death in 1374,[19] he was buried in the church of Santa Trinita, which was located opposite the school.

His writings range from mathematics to various other scientific subjects. He produced several treatises in the field of astronomy and also compiled tables: *Tabulae planetarum ad annum 1366*. His work was certainly highly appreciated because he was praised by both Giovanni Villani and Giovanni Boccaccio. Filippo Villani, Giovanni's nephew, places Paolo Dagomari among the most celebrated Florentines, particularly for his activity as an astronomer. Of him, Filippo says, "He was the greatest and most expert in geometry and arithmetic and exceeded all other ancient and modern astronomers." According to Villani, he was also the first to compose almanacs.[20] His merits earned him an election among the priors of Florence in 1363.

---

[16] Other historians point to him as a member of the Ficozzi family.

[17] The treatise of Blaise the Elder presents one hundred and fourteen problems, twenty-eight of which are of a practical nature. Special cases of third- and fourth-degree equations are also addressed in the treatise; see Ambrosetti, *op cit.*, p. 253.

[18] See Paolo dell'Abbaco, *Trattato d'aritmetica. Secondo la lezione del Codice Magliabechiano XI, 86 della Biblioteca Nazionale di Firenze*, edited by Gino Arrighi, Domus Galilaeana, Pisa, 1964, pp. 7–8.

[19] According to historian Nadia Ambrosetti, Paolo dell'Abbaco was born in 1288 and died in 1367, but this hypothesis seems to be contradicted by the reference to Paolo that Giovanni Boccaccio makes in his *Genealogie deorum gentili libri*, composed in 1373, where he mentions that Paolo dell'Abbaco was still alive.

[20] See Bartolomeo Veratti, *De' matematici italiani anteriori all'invenzione della stampa. Commentario storico*, Modena, 1840 (work dedicated to Prince Baldassarre Boncompagni).

He wrote two main works of a mathematical nature. They were written in vernacular and were both subsequently transcribed and republished in various editions under different titles: One is entitled *Regoluze del Maestro Pagholo astrolagho* and the other *Ragioni adatte a trafficho di merchatanzia* (it is also called *Trattato di tutta l'arte dell'abacho* or *Trattato d'aritmetica*).

The first text, which is part of a manuscript that is also devoted to astronomy, natural sciences, and medicine, consists of fifty-two rules expressed very succinctly in rhetorical style, without any demonstration, pertaining to arithmetic calculus and elementary geometry. For example, the first rule states that in a very large number, a dot should be inserted every three digits, from right to left, to separate thousands, thousands of thousands, and so on. The tenth rule states that to turn *lire* into *soldi*, you must multiply by 20. The fifteenth says that if you want to multiply two fractions, you have to multiply the numerators and denominators by each other. Other rules establish the relationships between different coins, still others concern the calculation of the phases of the moon, and yet more tackle geometric problems, such as calculating the circumference of a circle (for π, the Archimedean value $3 + 1/7$ is adopted, rule 32), calculating the volume of a well (rule 35), applying the Pythagorean theorem (rule 48), and so on.[21]

Treatise on Arithmetic, which was published in about 1340, is a more challenging work that somehow imitates *Liber abaci*. It contains both problems of an exquisitely mathematical nature and problems of applicative nature concerning mercantile calculus — in a sense, this might even be considered to be the main purpose of the work[22] — and finally problems of recreational mathematics. Compared to Fibonacci's *Liber abaci*, Paolo dell'Abbaco's text is very short and especially lacks the didactic ambition of the Pisan mathematician's work. In the version edited by Gino Arrighi, corresponding to the Codex Magliabechiano XI, 86, the text

---

[21] Paolo dell'Abbaco, *Le regoluzze di Maestro Paolo dell'Abbaco, matematico del secolo XIV*, Fondazione Datini, 1860.

[22] In one of the manuscripts, the text begins with these words: "In this book we will deal with various issues related to exchange of merchandise, drawn from arithmetic books and presented in vernacular by the excellent master Pagolo de Dagumari from Prato."

contains one hundred and ninety-seven problems, called "reasons", for each of which, in addition to the statement, a complete solution is provided, always in a rhetorical form. The exposition, which is enhanced by numerous illustrative drawings, has a didactic slant. However, while there is a certain gradualness in the way the different problems are presented according to their complexity (ranging from arithmetical operations on fractions at the beginning to solving second-degree equations toward the end), Paolo dell'Abbaco's work lacks a subdivision into well-defined chapters with precise didactic purposes, and the various topics are exposed in a rather disorganized manner.

The mathematical concepts are classic ones for this type of book. The book starts with operations between fractions and continues with proportions and applications of the rule of three (*reghola delle tre choxe*) as in the following example:

> Reason 43. A spear is beneath the ground 1/4 and 1/6 of all that it is long, and above the ground there are 20 arms; we want to know how long it is altogether. You should do as follows.

That is, the buried part corresponds to 10/24th of the length of the spear. So, the outer part corresponds to the 14/24th. The author goes on:

> Say it like this: for 14 I get 24, how much I will get for 20?

That is, calculate the proportion $14 : 24 = 20 : x$ from which we get the length of the spear $x = 34 + 2/7$ arms.

Next, there are problems traceable to first-degree equations that are solved by the false position or double false position method (which is curiously called *chatuino*, a term reminiscent of the Arabic term used by Leonardo Pisano, *elchataym*). This is followed by problems consisting of the sum of arithmetic and geometric progressions.

The work is also abundant with geometrical problems: the calculation of rectangular and circular surfaces (of course, in these cases, too, the value $3 + 1/7$ is adopted for π), and that of volumes (spheres, cylindrical wells, parallelepipeds, pyramids, etc.). Judging by the complexity of the problems of a geometric nature, the author's ability in this field would seem to be clearly superior to that of Leonardo Pisano. Several problems, for example, concern calculating the area of a circle section obtained by cutting the circle with a chord, calculating the area of the relative arc of a circle, calculating the sagitta, calculating areas of rectangles or triangles inscribed in such circles, and calculating areas of concentric circular crowns.

Finally, a series of problems consisting of solving second-degree equations is presented. Starting with problem 174, the author says, *Fa' choxì sechondo la reghola de l'arcjbra* (Do like this according to the rules of algebra). The proposed problems concern all six canonical classes of second-degree equations defined by Leonardo Pisano (dating back to the work of al-Khwarizmi) to define which Paolo dell'Abbaco uses the same terminology used by Leonardo: "number" to denote the constant, "thing" (*choxa*) for the unknown, and "value" (*cienxo*) for the square of the unknown. For each problem, the solution is given. Consider the following example:

> Reason 177. Give me two numbers such that one is double the other and 3 more; and when we multiply one against the other, they make 90. How much will each one be? Do as follows.

The problem requires one to calculate the value of the two variables x and y such that $y = 2x + 3$ and $xy = 90$, and can be traced back to the solution of equation $2x^2 + 3x = 90$, which belongs to the fourth class of second-degree equations, "when the values and the things are equal to the number." The text continues with the solution obtained by calculating $x = \sqrt{(90/2 + 9/16)} - 3/4 = 6$ and $y = 15$.

From the point of view of applications, the problems that concern the daily lives of merchants Paolo Dagomari addresses are also among those examined by Leonardo Pisano: bartering of goods (such as wheat for barley, cloth for wool, or copper for tin), the calculation of simple and compound interest (a typical interest rate seems to be 3 *denari* for one *lira* per month, 3 *soldi* per year, corresponding to 15 per cent per annum), exchanges of currencies (for example, between coins of Florence, Pisa,

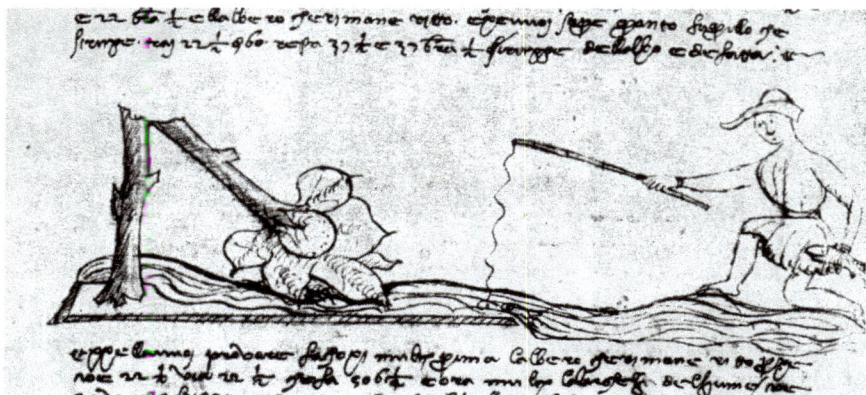

Figure 1. Reason 166 from Paolo dell'Abbaco's *Treatise on Arithmetic*. The problem requires the application of Pythagoras' theorem: a tree 60 fathoms high that stands on the side of a 30-fathom-wide river breaks, and the tip falls exactly across the river. The question is at what height the tree broke. Note that the same problem also appears in ancient Chinese texts as the "problem of the broken reed" (Paolo dell'Abbaco, *Trattato di aritmetica*, Codice Magliabechiano XI. 86, c 59v). (Reproduced by permission of the Ministry of Culture/National Central Library, Florence.)

and Lucca), divisions of profits among the members of a company, calculations of inheritances, or calculations of workers' wages.

Among the recreational problems proposed by Paolo dell'Abbaco, many belong to a repertoire frequently encountered in texts of this kind, even beginning with Alcuin's text or earlier Arabic and Chinese texts: for example, problem 42, the calculation of the number of trees that can be planted in a given piece of land; problem 50, the calculation of the speed a ship can reach with two sails, knowing the speed the two sails can raise if used separately; problem 66, decanting a liquid into given flasks[23]; problem 91, a dog chasing a fox; problem 108, an encounter between two travelers proceeding in opposite directions at different

---

[23] The problem is to obtain a flask with 4 ounces of liquid by having an 8-ounce full flask and two flasks, one of 5 ounces and one of 3 ounces, both empty. The problem, which is solved by an appropriate series of decantings, would also be taken up later by Luca Pacioli and was made famous by the movie *Die Hard 3*, in which neutralizing a bomb depends precisely on solving a similar problem.

speeds; problem 125, the contents of a purse; or problem 190, the bird problem.

For each proposed problem ("reason"), Paolo dell'Abbaco provides a precise solution algorithm, sometimes even indicating that there is more than one solution. In each problem, he says "Do as follows" to present the algorithm and in conclusion he says "and it is done and it is well". Sometimes, a solution concludes with the religious term "Amen", which is rather curious for a book on mathematics.

Before concluding, let us look at an example in which two different solution algorithms are proposed that lead to two different results, one of which is more correct, however. The example — problem 92, on renting a house to a number of tenants — highlights some subtlety of reasoning on the part of the author:

Reason 92. One rents his house to a master for 12 *lire* a year.

Each month, the house is inhabited by one more tenant, and when it comes to paying the rent at the end of the year, one has to calculate how to divide the expenses among the tenants.

The first tenant proposes the most trivial solution: Add up the total months of use of the house $(12 + 11 + \cdots + 2 + 1 = 78)$ and divide the total expense by this number. This results in everyone having to pay 3 *soldi* and 12/13 *denari* for each month they used the house. The other tenants, however, do not agree:

Don't do like this because you would cheat us, but let's do it as follows: let's pay this rent month by month. You will pay for the first month when you were alone, which comes to 20 *soldi*, then you and the one who came in the second month will pay 10 *soldi* each, and we go on like this until finally in the last month each will pay 20/12 *soldi*; and like this was done.

The Merchant's Workshop

In order to understand all the aspects and implications that make the art of calculating so crucial in the life of the medieval merchant, it may be useful to take a closer look at how merchant activities were carried out in the

14th century and how the notions learned in the abacus schools were used in daily practice. For this purpose, naturally, the primary sources are the archives of the merchants themselves, where they are available. One typical example is the archive left by Francesco di Marco Datini, a merchant who was born in 1335 and died in 1410. He worked in Prato and left a hundred and fifty thousand texts covering his entire professional life: commercial documents, letters and accounting records, and a wealth of information on goods, markets, prices, tariffs, duties, and currencies. The archive, which was only discovered in the 19th century, lets us understand the day-to-day management of merchant trade in Tuscany at the turn of the 14th and 15th centuries in great detail.

Alongside these collections of documents, an equally important role is played by the so-called *trattati di mercatura* (mercantile treatises), the manuals in which the necessary data for the mercantile calculations were provided.[24] After his period of schooling, in fact, a merchant had to call upon what he had learned to untangle the forest of coins from different cities and countries, different units of measurement of length, weight, and capacity, various methods of calculating debts and interest, and so on. Much information was acquired and noted down during merchants' travels and exchanges with other merchants, but the mercantile manuals were an essential working tool for quick access to the data needed to make valuations and calculations, and thus assumed an important role in a merchant's store, along with the abacus treatises that were probably still consulted from time to time.

Manuals relating to mercantile practice developed in Italy in the 13th and 14th centuries, but it is likely that Italian merchants had understood the usefulness of this material since the previous century, thanks to Arabic works such as the one by al-Dimisqui entitled *The Book of the Beauties of Trade and the Knowledge of Good and Bad Goods and the Counterfeits of Them that Fraudster Commit.*[25]

To get an idea of the contents of mercantile manuals, we can look at the one compiled in the first half of the 14th century by Pegolotti, which

---

[24] Sapori, *op cit.* pp. 17–19.

[25] Michael E. Tigar, *Law and the Rise of Capitalism*, Monthly Review Press, New York, NY, 2000, p. 77

is certainly one of the most complete and well-known manuals. Francesco di Balduccio Pegolotti (1290–1347) was a merchant and businessman employed by the Compagnia dei Bardi, a major Florentine trading company. Trading companies — which were generally based on the capital of a single family but open to contributions from various partners[26] — carried out financial activities that took the form of loans to individual merchants, other commercial enterprises, or, indeed, the governments of countries that needed substantial resources to undertake public works or finance wars. From a certain point of view, the Italian trading companies of that period could be likened to modern multinationals because they owned agencies in a variety of countries, their capital was transferred from one state to another through the mechanism of letters of exchange, and they could take autonomous political initiatives *vis-à-vis* Italian cities or the governments of other European countries, regardless of the interests of their home city. As we have already mentioned, the Compagnia dei Bardi had a large number of agencies in various European nations; in around 1340, it had three hundred and thirty-six employees.[27] Francesco Pegolotti operated in Antwerp in the service of the company from 1315 to 1317, in London from 1317 to 1321, and in Cyprus from 1324 to 1327, and in 1335, we find him involved in negotiations with the king of Armenia. At various times, and especially in the last phase of his life, he took on political roles in Florence, and when the Compagnia dei Bardi went bankrupt in 1347 due to the king of England's failure to repay a loan, he was involved in managing the crisis.

The original title of the text that is known today as *La pratica della mercatura* was *Libro di divisamenti di paesi e di misure di mercatanzie e d'altre cose bisognevoli di sapere a mercatanti*. The date when the work was written is a matter of debate. References to coins are an easy way of dating a manual like this because the presence or absence of

---

[26] A document drafted by a member of the Peruzzi family of Florence illustrates the creation of the trading company of the same name formed in 1324, in which seventeen members participated, nine from the Peruzzi family, who invested a total of 60,000 gold florins, each committing between a maximum of 5,500 and a minimum of 2,000 florins; see Sapori, *op cit.*, pp. 100–101.

[27] Francesco Balducci Pegolotti, *La pratica della mercatura*, edited by Allan Evans, in The Mediaeval Academy of America, 24, Cambridge, 1936, p. XVI.

references to specific coins that were first issued on a specific date provides unquestionable evidence. Generally, it is said that *La pratica della mercatura* was written between 1335 and 1342, but in reality, one must assume that a work such as this, which was a daily working tool for a mercantile company, was in a state of continual evolution, and one can therefore imagine that in any case earlier versions produced by Pegolotti himself or others were available in the offices of the Compagnia dei Bardi before 1335.

The work, which we know from a 1471 transcription, presents a long series of themes that were evidently of primary interest to a mid-14th-century merchant. The beginning is paradigmatic. It consists of short verses describing how a merchant should behave: "Integrity always suits him, Long foresight is good for him, And what he promises should not be lacking." It is interesting to note that the last of these verses states that the merchant must be able "To write reason well and not to err": that is, to do bookkeeping correctly.[28]

After providing the terminology used in the various countries to denote customs duties, trades (sellers, matchmakers, interpreters), units of measurement, and a variety of other mercantile terms (ships' freight, fairs, porters' fees, etc.), Pegolotti's encyclopedic work continues with a presentation of the key stages of the journey from the Black Sea to Catai (i.e. China), illustrating the goods to be found and the customs duties to be paid, if any, for each of the cities and nations encountered. The main body of the text consists of a review of more than forty cities and countries of the Mediterranean and Atlantic, from Constantinople to Sicily, from Morocco to Provence, and from Paris to London.[29] For each place, goods are listed that can be traded with their prices and the value they may have when sold elsewhere, the customs that may be paid (inbound

---

[28] *Ibid.*, p. 20.

[29] The list of places covered is very interesting for the purposes of understanding the breadth of the area in which Italian merchants operated in this historical period: Constantinople, Pera, Turkey, Armenia, Acre, Alexandria, Cyprus, Candia (Crete), Sicily, Chiarenza (Morea), Majorca, Tunis, Tripoli, Djerba, Venice, Friuli, Ancona, Puglia, Salerno, Naples, Gaeta, Florence, Pisa, Genoa, Nimes, Montpellier, Aigues-Mortes, Ibiza, Burgundy, Champagne, Paris, Flanders, Bruges, Brabant, Antwerp, London, Scotland, La Rochelle (Gascony), Seville, Safi, Salé, and Arzila (Morocco).

and outbound and differentiated according to the merchants' countries of origin; for example, in Sicily, the Genoese, Catalans, and Narbonnese did not pay on arrival, but if they bought goods, they had to pay a tax on leaving Sicily), the currencies in use and their rate of exchange, and the calendar of fairs. All this information is interspersed with tables of currency exchange or unit conversion. In addition, analytical data are also provided for some places on how the cost of producing goods (for instance, the cost of soap in Rhodes, oil and cheese in Puglia, or wool in England), the cost of extracting products (such as salt in Sardinia), or even the cost of services (such as unloading goods in the port of Genoa) is reached.

The text continues with long lists of spices, silks, fabrics, skins, furs, pearls, and precious stones, and the criteria for their valuation. Tables are then provided for calculating interest (from 1 to 8 per cent) compounded per 100 lire over a twenty-year period. Also included among the valuable information for merchants is advice on chartering ships. One part of the text that derives directly from the ancient *computus* tables is that relating to the lunar calendar: algorithms for calculating the date of the Easter Resurrection and tables of dates from 1340 to 1465, a calculation of the phases of the moon and the phase shift between solar and lunar years, and a determination of the constellation in which the moon is located at various times of the year. In addition, a chapter that shows how widespread the use of the abacus still was concerns suggestions for making calculations using a few *quarteruoli* (tokens).

Several parts of Pegolotti's work deal with alloys, particularly silver and gold alloys. In one of them, all the gold coins in circulation are listed, indicating their different carat contents, from 24 carat (florins and ducats) to 11 carat and similarly for silver coins. Next, silver alloys that come in pieces or rods are examined, and finally the alloys with which low-value coins are minted are listed.

Subsequently, the text contains some very sophisticated content that denotes a not insignificant level of technological expertise on Pegolotti's part: techniques for refining gold (to obtain purer gold, that is), for separating gold from silver, for testing raw silver, and, even more specifically, for assessing the amount of silver in an alloy.

What makes this part particularly interesting, with respect to the considerations we have offered about the relationship between mercantile treatises and abacus treatises, is the fact that it also presents algorithms similar to those proposed by Leonardo Pisano and Paolo dell'Abbacc for making alloys of a tenor (silver or gold) based on the availability of coins of different tenors. In all, we find fourteen problems (again called "reasons") on how to obtain silver alloys and six problems related to gold alloys.

One example of this will suffice here[30]:

I have 3 sorts of gold coins — that is 10 marks of 12 carats, 6 marks of 14 carats, and 5 marks of 18 carats — and I want to make an alloy by adding 22 carat gold and I want to add so much gold to obtain an alloy with 20 carats; I want to know how much 22 carat gold I need to add to my 3 sorts of coins; you should do as follows.

The solution algorithm proposed by Pegolotti consists of first calculating the average gold content of the 21 coins, namely, 14 carats, then calculating how many carats it is necessary to add because the 21 coins have 20 carats of gold each, and you get 126. To achieve this result, we must then recover 2 carats from 126/2 = 63 coins of 22 carats. In total, we will therefore have to alloy 21 + 63 = 84 gold marks of 20 carats each "as you required us to do."

As we have stated, Pegolotti's work has encyclopedic characteristics. It is not known whether he was aided by scribes when drafting the text or making the calculations necessary to compose the numerous tables. Certainly, he took advantage of the exceptional materials that must have been available in the offices of the Compagnia dei Bardi (or the materials to which he probably had access in the offices of the related companies of Peruzzi and Acciaiuoli), in particular regarding customs duties and fees of the various cities and countries, lists of suppliers of products (such as the long lists of English and Scottish wool producers), and the rates charged in the various ports. In addition, he must have had previous manuals at his disposal, such as *Memoria de tucte le mercantie come carican le navi in Alexandria*, which was compiled in Pisa in 1279, or almost contemporary ones such as the Venetian work *Tarifa zoe noticia dy pexi emexure di luogi*

---

[30]Pegolotti, *La pratica della mercatura*, pp. 354–355.

e tere che s'adovra marcadantia per el mondo.[31] In any case, Pegolotti's work has become a milestone in mercantile treatises and is the key reference internationally for historians dealing with these issues.

Various other writings of the same kind were produced up to the second half of the 15th century and contribute to our knowledge of the lives of medieval merchants and the role computational skills played in them. Among the most interesting of the works are *Pratica di mercatura* by Giovanni Acciaiuoli, a member of the mercantile company of the same name, and *Manuale di mercatura* by Florentine Saminiato de' Ricci, both from the late 14th century. Both Pegolotti's and Saminiato's works were widely taken up in later times by Giovanni di Bernardo da Uzzano (1420–1445). While heavily inspired by earlier works, his own work, *Libro di gabelle e pesi e misure di più e diversi luoghi e come pesi e misure tornano di un luogo ad un altro* (Book of duties and weights and measures from many various places and how weights and measures change from place to place), which he wrote in 1442, also reflects experiences formed in the family mercantile environment, such as concrete examples of accounts, expenses, and currency exchange.

Another equally interesting text is the one by Benedetto Cotrugli (1416–1469), who was born in Dalmatia in Ragusa (now Dubrovnik), and who operated in the Venetian sphere. This work, which is in four volumes, is entitled *Libro dell'arte della mercatura edito per Benedecto di Cotrugli ad Francisco di Stephano* (or *Della mercatura e del mercante perfetto*). It is dated to 1458 and contains both general indications on the technical skills a merchant should have and a large set of data on daily accounting. The first point we should underline relates to the knowledge the merchant needs to have (Book III, Chapter III: *De la sciencia de lo mercante*). Obviously, the merchant must be good at using the abacus, but he must also be literate: "And I say that the merchant has to be not only good at writing and using the abacus, as we said, but he also needs to be good with letters and rhetoric, because this is very useful since grammar makes a man clever and able to understand contracts and merchants make contracts every day."

---

[31] *Ibid,* pp. XXVII–XXVIII and p. XLVI. Because the Venetian treatise and Pegolotti's treatise were written at about the same time, historians cannot determine with certainty which of the two works borrowed content from the other.

One of the aspects on which the author dwells the most is the keeping of account books. Cotrugli suggests keeping three books: a "memoir", which today would be called the "first note", with an indication of the day's transactions; a "journal", with the accounting entries for each individual day; and a "notebook" (now called a "ledger"), with all the accounting organized for each customer or supplier. In particular, in Chapter XLI of Book I (*De l'hordene de tenere le scripture con ordine mercantile*). Cotrugli illustrates how the double entry should be organized: "*Item* every entry has to be written on both sides of the page: that is, what has to be given on the right, and what has to be received on the left. And for every entry you should note when, how much, to whom and why: when, that is the date, how much, that is the amount of money, to whom, that is whether he has to give or receive, and why, that is the reason for the transaction."[32]

Of course, the author stresses the need for the merchant to know how to make calculations correctly, in particular regarding the exchange of coins (Book I, Chapter XI): "And every calculation of yours should make reference to the reason for the exchange and to the amount of money that is exchanged for one ducat. For example, if I am in Barcelona and I wish to send clothes to Naples, I need to know how many ducats a bolt of fabric is worth in Naples; the principle and basis of your consideration has to be the exchange rate."[33]

The last work we feel it appropriate to mention is the one attributed to Giorgio di Lorenzo Chiarini, *Libro che tracta di mercatantie et usanze de' paesi*, the oldest manuscript of which is dated to 1458. It consists mainly of an extension of earlier documents, much of which is taken from Pegolotti's text. The manual begins with these words: "Here begins a book of all customs, exchanges, currencies, weights, and measures, and about letters of exchange as they are used in various countries and various lands."[34]

---

[32] Benedetto Cotrugli, *Libro dell'arte della mercatura*, edited by Vera Ribaudo, Edizioni Ca'Foscari, Venice, 2016, p. 83 (https://phaidra.cab.unipd.it/detail/o:432210?mycoll=o: 432598).

[33] *Ibid.*, p. 76.

[34] Travaini, *Monete, mercanti e matematica*, p. 72.

Since Chiarini himself claimed to have copied parts of the text in Ragusa in the environment of Tuscan merchants, various historians prefer to define the author as "pseudo-Chiarini."[35] The work was later incorporated into Luca Pacioli's *Summa de arithmetica*, and for that reason, it acquired a certain fame. Many parts of Chiarini's and Pegolotti's texts "run closely in parallel,"[36] although at various points, Chiarini adds additional information and details to those provided by Pegolotti.

As mentioned earlier, when one reads mercantile manuals, one realizes the vast amount of information that must have been available in the stores of merchants, especially those belonging to the large merchant and banking companies such as Bardi, Peruzzi, and Acciauoli. Historian Evans imagines tables of data posted on the walls of the offices. This information had to be updated continuously in order to take into account changes in prices, tariffs, and duties, and the fluctuating exchange rates. The various mercantile treatises list hundreds of coins minted in the different cities and nations[37] of Europe and the entire Mediterranean area, with their respective exchange rates. One can understand how the calculation skills of merchants were put to the test by such a wealth of data and the valuations needed to make the activity profitable, profit clearly being the merchant's ultimate goal, obvious though this may seem.[38]

Although many merchants had acquired skills in numerical computation based on Indo-Arabic numerals, it seems that in several merchant stores in various European countries, there continued to be an instrument for bookkeeping that we might imagine would have fallen into disuse by this time: the abacus (Figure 2).

In this historical period, the abacus still had an important role in the large offices of merchant companies. It had imposing dimensions: "a plane of about 3.30 × 1.65 meters belted by an 8 centimeter slat and covered with a dark-colored cloth, on which, in chalk or otherwise, were

---

[35] *Ibid*, p. 164.

[36] Pegolotti, *op cit.*, p. XLIV.

[37] Uzzano alone mentions more than a hundred and ten gold, silver, and low-grade coins. Travaini's text, *op cit.*, extensively reports all the lists of coins that are mentioned in the mercantile treatises and account books of Francesco Datini and other merchants.

[38] It is said that Francesco Datini's motto was "In the name of God and profit."

Figure 2. Merchant and clerk using an abacus with tokens (Engraving on copper from 17th century — Bridgeman Images).

drawn 6 and 4 parallel straight lines vertically and horizontally respectively that by intersecting each other formed 35 quadrilaterals."[39] The tokens used to mark the values in play, which were called *quarteruoli* in Tuscan, represented *denari* in the first column on the right, *soldi* in the second, *lire* in the third, tens of *lire* in the fourth, hundreds of *lire* in the fifth, thousands of *lire* in the sixth, and tens of thousands of *lire* in the seventh. The horizontal rows were intended to contain the representation of the operands of sums, multiplications, and so on. In England, the same counter was called the "exchequer", and it is interesting to note that the division of monetary values was quite similar: 12 pence for a shilling, and 20 shillings for a pound. Evidence of the use of the abacus in merchants'

---

[39] Sapori, *op cit.*. pp. 110–111.

(a)   (b)

Figure 3. A brass abacus token of German manufacture. The token bears the name of the manufacturer, Hans Schultes (who was active in Nuremberg in around the middle of the 16th century), on one side (a) and an image of the lion of St. Mark on the other (b) (from which it can be assumed that it was intended for Venetian customers).

stores in the 14th century is again provided to us by the merchants' own accounting books. In the Peruzzi company's records, we read, "lbr. 212 s. 2d. 10 a fior., for writing papers and books [...] lbr. 20, s. 11 1/2 a fior. for lbr. 61 ounces 5 of *quarteruoli*."[40] Of course, the division into rows and columns could vary depending on the size of the figures in play (which could also reach millions of florins). Thus, the tokens used for calculations could also take on different values (tokens of different colors or sizes could indicate the unit or value 5 or value 10, or differentiated values could be represented by placing the tokens at different points of the quadrilaterals (bottom, top right, or top left) (Figure 3).

An alternative and far more manageable calculation medium, the use of which is illustrated by Pegolotti,[41] consists of small tablets in which boxes are given for all individual values 1–12 *denari*, 1–20 *soldi*, 1–10 *lire*, 20–100 *lire* (ten by ten), and 200–1,000 *lire* (one hundred by one hundred). In this case, the tokens are arranged in the boxes in an additive way. For example, to represent the value 879 *lire*, s. 10, d. 5, the tokens

---

[40] *Ibid.*, p. 111.
[41] *Ibid.*, p. 113.

are placed on the boxes of 800 *lire*, 70 *lire*, 9 *lire*, *soldi* 10, and *denarii* 5. Where the same amount is represented on the abacus with tokens of unit value, the saving is obvious: Only 5 tokens are used instead of $8 + 7 + 9 + 10 + 5 = 39$ tokens. Even in the case where tokens of unit value and tokens of value 5 were used on the abacus, $4 + 3 + 5 + 2 + 1 = 15$ tokens would still be necessary.

Chapter 7

Computing in Humanistic Culture

Merchant Arithmetic in Europe in the 15th Century

The production of abacus books intensified further in the 15th century, both in Italy and in other European countries. In Italy, the predominance of Tuscany gave way to a flowering of contributions published in various regions: Veneto, Emilia, Rome, and the South.[1] We will devote the following sections of this chapter to these developments; however, it seems important at this point to dwell on an interesting aspect of the 15th century with regard to abacus culture: namely, its spread to other European regions. As Kurt Vogel, one of the leading scholars on this subject, has noted, at the beginning of the century, the supremacy of Italian mathematical culture, particularly with reference to applications, was absolute. "German students studied at Italian universities, merchants' sons were pupils of the abacus masters of Venice. [...] In the terminology of merchant arithmetic and algebra the words in use were Italian — such as *cosa* (thing) to indicate the unknown of an equation. Likewise later German painters came to Italy to learn perspective."[2] It should also be kept in mind that Italian was the *lingua franca* of merchants in Europe and various parts of the Mediterranean area, which also facilitated the penetration of Italian mathematical works. Indeed, while abacus calculus had developed fairly

---

[1] Ambrosetti, *L'eredità arabo-islamica*, pp. 256–259.

[2] Carlo Maccagni (ed.), *Atti del primo convegno internazionale di ricognizione delle fonti per la storia della scienza italiana: secoli XIV-XVI*, Barbera Editore, Florence, 1967, p. 120.

early in Italy (in the late 12th to early 13th century) in relation to the needs of merchants' business activities, the development of mathematical knowledge was slower in other regions of Europe. For example, at first, the only few works that circulated in Germany along with Italian texts were mainly of French origin, such as those by Sacrobosco and Villedieu that we mentioned previously, as well as a few manuscripts that were translations of Arabic texts. It was only around the middle of the 15th century that German abacus books began to appear. The first known one is entitled *Algorismus Ratisbonensis* and contains classical problems of mercantile mathematics, alongside the six classical cases of second-degree equations defined by al-Khwarizmi. The text known as *Dresden Manuscript*, which dates from 1481, also contains the six canonical cases of second-degree equations, but to these are added eighteen more cases of higher degrees as well as numerous problems of both a commercial and a recreational nature.

One character who played an important role in the bond between the Italian and German mathematical cultures was Regiomontanus (Johannes Müller von Königsberg, 1436–1476), who after studying mathematics and astronomy in Leipzig and Vienna, traveled to Italy, where, under the protection of Cardinal Bessarione, a philosopher and humanist of Byzantine origin, he had access to an immense body of ancient and recent mathematical literature,[3] part of which he brought back with him to Germany. The humanists' interest in classical culture and the rediscovery of ancient works were not only important for literary culture but also had a significant spillover effect on mathematical culture. Regiomontanus was the author of a number of works on astronomy and geometry, such as *De triangulis omnimodis* and *Tabulae directionum*, which made an important contribution to trigonometry and which in part show his familiarity with Greek (Euclid) and Arabic (Nasir Eddin) sources. In 1474, Regiomontanus compiled a catalog of works from previous centuries and antiquity that he intended to have printed at his printing press in Nuremberg. Although the project was not successful, the cultural vision that emerged from it

---

[3]The size of Cardinal Bessarione's library is evidenced by the fact that his donation to the Republic of Venice was the initial act in the creation of the Biblioteca Marciana; cf. Ambrosetti, *op cit.*, p. 288.

strengthened his image. His short life did not allow him to delve deeper into the fields of arithmetic and algebra, but it should be acknowledged that he played a fundamental role because he acted as a liaison between the Italian humanistic world and the mathematical milieu of Central Europe by circulating Italian mathematical texts as well as translations of al-Khwarizmi's works.

France also reveals a significant gap from Italy in terms of the production of abacus literature. Compared to the hundreds of treatises published in Italy between the 13th and 15th centuries, there were only twenty-five works[4] in France during the same period. After the success of Sacrobosco's and Villedieu's books on "algorisms", which, as noted previously, had long been adopted and used in universities, there is no record of significant contributions related to abacus mathematics until the second half of the 15th century. The only remarkable exceptions are Johannes de Muris' work *Quadripartitum numerorum*, which was very much inspired by al-Khwarizmi's algebra, and the aforementioned abacus treatises by the Italians resident in Montpellier, Jacopo da Firenze, and Paolo Gherardi.

Among the works that were published in the second half of the 15th century, *Tripartitum numerorum* (*Triparty en la science des nombres*) by Nicolas Chuquet (1445–1488) played an important role. The text, which the historian of mathematics Carl Boyer considers to be "a work that in level and importance is perhaps the most remarkable after the *Liber abaci*," dates from 1484, but it was not published until the 19th century when it appeared in Prince Baldassarre Boncompagni's *Bulletin*. It consists of three parts: one devoted to calculus with integers and fractional numbers, one on calculus with radicals, and one devoted to algebra. There are numerous similarities with *Liber abaci*, and one might imagine that the author knew Fibonacci's work, or at least some of his transcriptions. In particular, the application problems take up the Italian tradition. The text begins with a presentation of Indo-Arabic numbers (including zero) and continues with classical concepts: progressions, proportions, the rules of simple three and compound three, and the false position and double false position methods. As mentioned earlier, the second part deals with calculus with radicals and root extraction, and the third with algebra.

---

[4] *Ibid.*, p. 260.

In this part, some aspects retain the traditional approach: For example, quadratic equations are solved by the ancient geometrical method of filling the square. Some of the new features introduced by Chuquet are interesting, however: First, there is the style of presentation, which is no longer just rhetorical but also contains elements of formalization. For example, to indicate the monomial $3x^k$, Chuquet writes 3^k. Also, for the first time, he uses negative numbers, which are denoted by an *m*, both as coefficients and as exponents, and he points out, also for the first time, that the exponent 0 assigned to a number makes it equal to 1. When writing m.12^{2m}, for example, Chuquet means $-12x^{-2}$. Another (less relevant) novelty is that he uses French terms such as *premiers* instead of *cosa* and *champs* instead of *census*. A final innovation in Chuquet's text is the introduction of the terms *million* (10^6), *billion* (10^{12}), and *trillion* (10^{18}), albeit with different meanings from the ones we use today.

The spread of arithmetic and algebraic treatises in the 15th century was not limited to the Germanic area and France, of course. Nadia Ambrosetti's book on the Arab-Islamic legacy in European mathematics presents an extensive review of works on calculus in various other areas of late medieval Europe. The survey goes from England, where translations of Villedieu's and Sacrobosco's works, known as *The Art of Nombrynge* and *The Craft of Nombrynge*, respectively, were made in the 15th century, to the Iberian peninsula, where two manuscripts, *De Arismetica* and *El Arte del Alguarismo,* were printed. In particular the former of these two texts had the typical contents of abacus books while the latter was almost exclusively devoted to metal alloys. Again in the Iberian peninsula, in 1482, the first trade treatise text in Catalan, *La Summa de la art de Arismetica* by Francesc de Sanctcliment, was printed. Finally Ambrosetti's survey also takes into account works produced in the Greek-Byzantine area and the Jewish community operating in the Middle East.

Italian Abacus Masters in the 15th Century

Returning to the Italian sphere, we see that the production of abacus books did not decrease but rather intensified throughout the 15th century. Of the two hundred and twenty abacus treatises and texts on elementary geometry produced in Italy between 1276 and 1500, and mentioned in the previous chapter, more than half were produced in the second part of the 15th century.[5] Generally, the treatises published during this period retained the classical approach in their mathematical content, in the applications of mercantile interest, and, finally, in the problems of recreational mathematics. The only innovations concerned the algebra or geometry sections.

In the Tuscan context, some importance is generally attributed to Maestro Benedetto di Antonio da Firenze (1432–1479), the author of *Trattato d'abacho* published in 1459 and *Trattato di praticha d'arismetrica* published in 1463.[6] Large parts of the second work are in the form of a compilation, and so it offers an interesting summary of the teachings of the abacus schools of the late 14th to early 15th centuries. The text consists of sixteen books, some of which, as the author states, quote texts by earlier authors, beginning with the vernacular version of excerpts from the Latin translation of al-Khwarizmi's algebra prepared by Guglielmo di Luni in the 13th century. Furthermore, in the fourteenth book of *Treatise*, the author reports a series of a hundred and fourteen algebraic problems, twenty-eight of which are of a practical type and relate to the mercantile sphere, and are taken from the work of Biagio the Elder (who, as noted previously, had been one of Paolo dell'Abbaco's teachers). In the fifteenth book, he presents problems taken from the works of Leonardo Pisano, Giovanni di Bartolo, Antonio de' Mazzinghi, and other 14th-century abacus masters. Finally, the sixteenth book contains the text of Leonardo Pisano's *Liber quadratorum* with some additions from the hand of Master Benedetto. However, the algebraic part of the work is of some interest and contains several original themes.[7]

In the second chapter of the thirteenth book, Benedetto uses numerical examples to state several rules for transforming expressions containing radicals — such as the one that when expressed in modern terms, states that $(\sqrt{a} + \sqrt{b})cx = \sqrt{(ac^2x^2)} + \sqrt{(bc^2x^{2})}$. In the third chapter, in addition to

[5] Høyrup, *Leonardo Fibonacci and* Abbaco *Culture*, p. 35.

[6] It seems that *Tratato di praticha di geometria*, the author of which is uncertain, could be added to these works; see Raffaella Franci and Laura Toti Rigatelli, *Maestro Benedetto da Firenze e la storia dell'Algebra*, in "Historia Mathematica" 10, 1983, pp. 297–317.

[7] *Ibid.*, pp. 301–304.

the six canonical classes, he presents no fewer than thirty-six other classes of equations of increasingly higher degrees, from the second to the sixth, providing a rule for solving each one that often coincides with a reduction of the equation to an equation of the second degree. For example, the last class considered is that of equations of the type $x^6 = bx^5 + cx^4$ whose non-zero solution is simply $x = b/2 + \sqrt{[(b/2)^2 + c]}$.

Another work that deserves attention is the one by Pier Maria Calandri (1456–1508). As already mentioned, the Calandri family was an important family of abacus masters in the 15th century. The craft was inherited, so to speak, by Master Luca, the father of Checca, Piero Calandri's wife, and was continued by their son Maestro Calandro di Piero di Mariano, who had the aforementioned Maestro Benedetto da Firenze as his pupil, and by his grandsons Maestro Pier Maria and Maestro Filippo Maria.[8] In a cadastral document from 1480, Pier Maria himself declares, "Io Pier Maria fo l'abacho tra' Pilliccai" (I myself Pier Maria do the abacus among the furriers).[9] *Tractato d'abbacho* was written in around 1480.[10] The text is avowedly aimed at mercantile applications, as is evident in the *incipit*: "*Incomincia uno tractato d'abbacho nel quale si dimostra quello che s'appartiene alla mercatantia fatto da Piero Maria a un suo amicho*" (Here begins an abacus treatise in which it is shown what belongs to the merchant business carried out by Pier Maria for one of his friends). The introduction expresses the book's primary interest thus: "Among all the mathematical faculties which by their certitude are known, none is found to be more playful than arithmetic; but the part which concerns the merchant by far advances the others in its usefulness."

The work consists of twenty-three chapters, eight of which are devoted to "delightful cases" since, as the author says in Chapter 6, "every sound intellect would be in annoyance not reasoning about other cases than merchants affairs. Therefore [...] I mean to show some

---

[8] Filippo Maria Calandri was also the author of a mathematical work, *Arithmetica* written in 1491 for Giuliano de' Medici, son of Lorenzo the Magnificent; see Franci (ed.), *Alcuino di York*, p. 152.

[9] Calandri, *Tractato d'abbacho*, p. 12.

[10] According to Sergio Calzolani, the text is not the work of Pier Maria Calandri, but was instead written in 1459 by Maestro Benedetto da Firenze (www.geometriapratica.it/index.php/it/11-geometria-pratica/49.piermariacalandri).

delightful cases." At the beginning of the work, two ways of performing the product of two numbers are presented in artistic boxes: 99,999 by 44,444. The first method is called "multiplication by square" and corresponds to the method that is sometimes called *per gelosia* and in Leonardo Pisano's *Liber Abbaci* is called *in forma scacherii* (in which the partial products are written in columns but are summed diagonally); the second method is called "multiplication by *beriquocolo*",[11] which corresponds to the method sometimes also called "by organ" or "by ladder" and is normally used by us today (in which the partial products are written diagonally and summed in column).

Right from the first chapter, the text goes into topics of mercantile interest, with problems concerning the cost of commodities such as grain, wine, eggs, and chickens. The arithmetic techniques used are from the classical repertoire: arithmetic operations between fractions, the rules of three simple and three compound, calculations of simple and compound interest, and so on. Many are taken from earlier texts such as those by Master Benedetto da Firenze, or even go back to Leonardo Pisano. Conversely, algebraic problems regarding second-degree equations are entirely absent. The application topics are also entirely traditional: the buying and selling of goods, bartering (typically wool for cloth), divisions of profits and losses between partners in a company, metal alloys, exchanges of coins, and units of measurement. Chapter 7 is very interesting in this regard because of the glimpse it gives into the European mercantile marketplaces in which Florentine merchants carried out their business activities, in which the monetary exchanges among various Italian cities (Venice, Bologna, Milan, Genoa, and Palermo) and foreign cites (Valencia, Barcelona, Avignon, Geneva, Bruges, and London) are illustrated.

The problems of the "delightful" type also fall into a traditional groove that goes back to Leonardo Pisano and even earlier recreational mathematics: chasing problems, calculating numbers by the remainder of divisions, ships sailing with one, two, or three sails, men buying horses, purse problems, the height of partly buried masts, tiling a floor, digging a

---

[11] Beriquocolo is an ancient Florentine cake made of flour and honey.

well, and so on. One interesting, more original problem that appears in Chapter 15 is the so-called *Joseph problem*:[12]

> The tale is about 15 Christians and 15 Jews who were on a ship and after a storm they agreed that 15 of them should be thrown overboard to lighten the ship. And they made a circle, then they gathered round, and those who touched 9 were thrown into the sea. Whereupon it happened that one of those Christians made it in such a way that all those Jews were thrown into the sea and the Christians remained safe.

The solution is presented with an illustration of the disposition of the Jews and Christians that depicts what was (maliciously) planned by one of the ship's passengers: Counting from 9 to 9, only the Christian passengers survive, and all the others are thrown overboard.

Chapter 23 of *Tractato d'abbacho* is devoted to geometry. It begins with a reference to truth: "Geometry by figures seeks the truth of things." This could be interpreted as a recognition of the important role in geometry of the ability to define abstract concepts (such as points, lines, surfaces, and angles) and to determine the properties of geometric figures using an axiomatic-deductive approach. The chapter contains a series of definitions and a number of problems concerning plane figures such as triangles, circles, and solid figures (for instance, wells, canals, columns, pyramids, and spheres). Of course, there is no shortage of classic applications of the Pythagorean theorem, such as those concerning a rope that starts from the top of a tower and reaches beyond a river flowing under the tower or that of the broken tree whose top falls to the ground some distance from the base of the tree. Pier Maria Calandri's interest in geometry is evidenced by the fact that he wrote another work on geometry, *Compendium de Agrorum corporumque dimensione*, in four chapters of a more applied nature. Indeed, the work begins thus, "Geometry can be divided into two parts, of which one is called theoretical and the other practical, but the theoretical we shall leave to

---

[12] The problem is so called because its origin lies in an episode involving a certain Joseph, which was narrated in *De bello giudaico* by Josephus Flavius. The same problem is presented in the works of Filippo Maria Calandri; see Franci (ed.), *op cit.*, pp. 151–154.

philosophers, and of the practical, related to numbers that for measuring the earth are needed, we shall say."

As mentioned earlier, by the 15th century, the production of treatises on calculus was no longer a monopoly of the Tuscan mathematical community but had spread to other regions as well. In particular, the work entitled *L'arte de l'abbacho* (also known as *Aritmetica di Treviso*) was written in the Venetian language by an unknown author and was printed in 1478 in the Veneto region. It went down in history as the world's first printed mathematical text; Euclid's *Elements* was printed in Venice only four years later, in 1482.

This is a practical manual designed for everyday use by merchants, and like the treatises we have already encountered, it concerns the arithmetic of commercial transactions.[13] First comes an introductory part in which the Indo-Arabic figures are presented. Among the figures, "the first figure, that is 1, is not called a number but is the beginning of all numbers. And the tenth figure, that is 0, is called 'cipher of nothing', that is figure of nothing because by itself it does not mean anything but juxtaposed to other figures makes their value grow."

Then, after the fundamental arithmetic operations have been introduced, the text is devoted to typical problems of mercantile activity and the algorithms for solving them. As for arithmetic operations — "Five are the acts: that is, enumerate, sum up, subtract, multiply and divide" — the author does not present any method other than those practiced at the time, except for subtraction. In fact, the interesting idea is that when the digits of the subtrahend are greater than the digits of the minuend, instead of borrowing between the digits of the minuend, one can add one to the digits of the subtrahend. Consider the example $1,004 - 826 = 178$. In the first step, we should calculate $14 - 6 = 8$, then we should remember that the second digit of the minuend became 9 instead of 0 and we should calculate $9 - 2$. In the method suggested by the work we are examining, instead, it is proposed that the second digit of the result should be determined by adding 1

---

[13] Anonymous, *L'arte de l'abbacho*, 1478 (www.centromorin.it/info/abacho/abbacho.pdf). See also Frank J. Swetz, *Capitalism and Arithmetic: The New Math of the 15th Century*, Open Court, La Salle (IL) 1987.

to the second digit of the subtrahend and calculating $10 - 3 = 7$, and so on, avoiding the sequence of borrowing and making a simpler calculation.[14]

The first topic of practical interest the unknown author deals with is the conversion from *lire* to *soldi*, with related multiplication tables for 20, and the calculation of carats of gold coins, with related multiplication tables for 24, 32, and 36. Next, following the classical pattern, come applications of the rule of three things — that is, the calculation of proportions in the purchase and sale of goods (wheat, pepper, cloth, and "French" wool). Another typical topic covered is that of company formation:

> Three merchants, Piero, Polo and Zuanne formed a company. Piero invested 112 ducats, Polo invested 200 ducats, and Zuanne invested 142 ducats. Their activity brought a profit of 563 ducats. I ask what should go to each one of them.

The total investment amounts to 454 ducats, so the problem is brought down to a calculation of proportions. If x is the profit that goes to Piero, we have (with modern expression) $454 : 563 = 112 : x$, and so on for the other partners. This is followed by a series of other problems such as: the formation of companies in which, rather than capital, the partners pool goods of various prices, problems of bartering (for instance, balm for wax, sugar, and ginger), problems concerning two travelers going in opposite directions meeting at different speeds, chasing problems, purse problems, problems concerning alloys, problems concerning building constructions with various numbers of workers, and so on. All in all, a completely classical abacus repertoire. Completely absent, on the other hand, are geometric and algebraic problems (equations of second degree).

Another interesting contribution from around the same time came from the Campania area and was produced by Pietro Paolo "Nolense" Muscarello. The text, which was entitled *Algorismus*,[15] was completed in 1478 and appears to have been dedicated to the Nolan family of Albertini.

---

[14] See Giorgio T. Bagni, *Larte de labbacho* (*The Treviso* Arithmetic, *1478*) *e la matematica medievale*, in *I Seminari dell'Umanesimo Latino 2001–2002*, Fondazione Cassamarca, Treviso, 2002 (www.syllogismos.it/history/uma-abbaco.pdf).

[15] Pietro Paolo Muscarello, *Algorismus. Trattato di aritmetica pratica e mercantile del secolo XV*, edited by Giorgio Chiarini, Banca Commerciale Italiana, Milano, 1972.

Information about the author's life, and his reasons for writing the text, is very scarce. A fleeting mention of a "young man" in the introduction to the work suggests to some scholars that the text was written for didactic purposes, in particular as a basis for preparing lessons.[16] On the other hand, as we shall see, the content of the manuscript is at times far from trivial and suggests that the work was not intended for beginners. As for the setting in which it was conceived, the fact that the city of Nola is mentioned several times in the work and that various goods and products from the province of Nola, such as hazelnuts, appear repeatedly in the problems presented would suggest that the book originated in Nola itself, but the mathematical validity of the treatise (despite the presence of several errors) and the richness of the miniatures that illustrate it lead one to believe that it actually originated in the cultural milieu of the Neapolitan court.[17] The text is enriched by numerous valuable illustrations, but unfortunately, even their genesis is unknown; they may have been made by the author of the manuscript himself, as would be suggested by the fact that they help with an understanding of the problems and that the author himself mentions them with the phrase "as you can see in the drawing". On the other hand, on the basis of a number of considerations regarding the incomplete correspondence between what is stated in the problems and what is portrayed, it can be assumed that the illustrator is someone other than the author.[18]

Muscarello's manuscript consists of one hundred and eleven sheets written on both sides. It is therefore fairly voluminous and presents a very long sequence of exercises and problems without a real division into chapters. The index at the beginning of the manuscript lists two hundred and twenty-five problems ranging from elementary operations on fractions to the resolution of complex high-degree equations. It presents problems belonging to the most classic repertoire of calculations of a mercantile nature (bartering, exchanges of coins, divisions of profits between partners of trading companies, etc.), of a geometric nature (applications of the Pythagorean theorem, calculation of surfaces and volumes, etc.), and of

---

[16] Mauro Nasti, *L'algorismo nell'aritmetica del Trecento e del Quattrocento*, in Muscarello, *op cit.*, p. 284.

[17] Ambrosetti, *op cit.*, p. 257.

[18] Maria Luisa Gengaro, *Le miniature del codice*, in Muscarello, *op cit.*, pp. 306–308.

recreational mathematics (chase problems, etc.). The solutions to all the problems are presented in rhetorical form, and counterevidence is also often provided. It is interesting to note, however, that the textual listing of operations to be performed is supported by diagrams in which numerical operations are presented with lines connecting operands, partial results, and final results. Unfortunately, as mentioned, the calculations performed and the solutions provided are incorrect in certain cases; sometimes, this is the fault of the copyist, but in some places, it seems that the error is to be attributed to the author himself.[19]

The philosophical premise of the text is presented in a kind of preface in which the author points to the two main virtues of wisdom and knowledge, identifying Solomon as "the first wise man in the world" and the sacred scriptures as the source of "so much science to our benefit." With regard to geometry, a further quotation that helps illustrate the cultural context of the work is the implicit reference to Euclid, when Muscarello states, "Similarly, geometry is a liberal art, which is the art of measure and with the help of this art we might measure the height up to the empyrean sky. And of this art the first geometer edifier is mentioned by Dante in the Book of Hell."

It is interesting to note that in the preface, Muscarello provides the (fake) etymology of the word *algorismus*: "And you should know that we call it Algorismus because this science was first started in Arabia and who invented it was likewise Arab; and art is said 'algo' in Arab and number is said 'risimus' and therefore is said Algorismus."

Furthermore, the algorithms are classified into nine fields: "And the scopes of the said Algorismus are nine, that is number, sum, subtraction, halving, doubling, multiplication, division, progression, calculation of square root."[20]

The treatise begins with a variety of tables: multiplications (the complete Pythagorean table up to 20×20 and multiplications of larger numbers, up to 53, by multiples of 10 up to 100 are presented), squares (up to the square of 100), and decomposition into factors and prime numbers (up

---

[19] Nasti, *op cit.*, pp. 297–300.

[20] As can be seen, Muscarello takes both the etymology of the word *algorismus* and the classification of arithmetic operations from Jacopo da Firenze's *Tractatus Algorismi*; see *ibid.*, p. 281.

to 30). Then, after the table of contents, the presentation of exercises and problems (which are again called "reasons") begins. The first parts are devoted to elementary operations with fractions (multiplication, addition, subtraction, and division); operations on integers are evidently taken for granted. Reduction to the least terms is frequently used, while to carry out sums and subtractions between fractions, the calculation of the lowest common denominator is not used but the denominators are just multiplied among themselves, postponing any simplifications to a later time.

The following section (*De fare ragione*) concerns proportions and the rule of three simple and three compound, which is called the "foundation of any merchant calculation." Indeed, this part is full of applications to the mercantile world with frequent references to the Nolan environment (buying and selling of goods, even in different locations, exchanges of coins, metal alloys, changes of units of measurement, etc.).

One example will suffice:

Make this reason: 10 arms of fabric from Verona correspond to the length of 13 arms in Naples and 14 arms of Naples correspond to 15 arms in the measure of Nola. I am asking to how many arms of Verona will correspond to 30 arms of Nola.

The solution — "so much will correspond in Verona to 30 arms of Nola, i.e., arms 21 + 7/13, and this is correct" — is obtained by applying the rule of three compound ("this requires the rule of three compound, with which you will be able to make many reasons").

This is followed by a section on solving first-degree equations using the false position method or double false position:

Example. There are three fellows that should share 25 ducats; I will not say how much goes to the first one. I will just say that the second will have twice the first one and the third one will have twice as many as the second one. I am asking what amount will go to each one without tricks.

Muscarello proposes various solution methods for problems of this nature using the simple or the double false position method. If we call x the amount that goes to the first fellow, assuming $x = 1$ gives a total of 7 instead of 25. Assuming $x = 4$ gives a total of 28 instead of 25. Computing the proportions $x : 1 = 25 : 7$ or $x : 4 = 25 : 28$ gives the desired result of $3 + 4/7$. In addition to the methods based on false position, Muscarello also proposes the method that directly gives the correct solution:

> Now I want to show a different approach which goes all at once.

In this case, 1, 2, and 4 are added together, obtaining 7. This is then the number of parts into which the total is to be divided. To determine the value of each part, it is sufficient to divide 25 by 7, still obtaining $3 + 4/7$. Needless to say, the latter method corresponds to the one we would use today to address the given problem by solving the first-degree equation $x + 2x + 4x = 25$ and obtaining the result $x = 3 + 4/7$.

The next sections of Muscarello's text deal with barters (*De baracti*) and with the calculation of interest on loans (*De imprunti*), with a somewhat traditional content. Many other parts of the text follow the classic pattern of abacus treatises. Also traditional are the inheritance problems, the problems of summing arithmetic successions, and some recreational problems (for example, look for a number that when divided by 2 gives remainder 1, divided by 3 gives remainder 2, ... , divided by 10 gives remainder 9, the solution of which is 2,519).

As far as geometry is concerned, the problems presented involve various applications of Pythagoras' theorem (such as the problem of the broken tree, which, we should remember, already appears in ancient Chinese texts), the calculation of surfaces and volumes (for example, the problem of calculating how the level of the liquid contained in a tank changes if a column of given dimensions is introduced into it), the calculation of circular areas and circular crowns, and so on.

The most interesting and original part, on the basis of which it can be argued that the work is at a higher mathematical level than similar works,[21] is the section devoted to calculations with radicals and the solution of particular classes of high-degree equations. Muscarello introduces six types of roots: the square root, the cube root, and the fourth root ("root of root") corresponding to notions familiar to us. In addition,

---

[21] *Ibid*, p. 288.

he introduces the "deaf" root, which corresponds to the non-integer square root ("the one for which one cannot give an integer solution as it might be the root of 12"), the *pronic* root of an integer n, that is, a value x such that $x^2 + \sqrt{x} = n$ (for example, the pronic root of 84 is 9), and the *relata* root of an integer n, a value x such that $x^2 \sqrt{x} = n$ (for example, the relata root of 32 is 4). At this point, the author envisages an increase in the difficulty of the problems presented: "step by step we go to numbers of more subtlety." In the first instance, it is shown how to perform the operations of multiplication, division, addition, and subtraction of radicals. Next, a series of problems are posed that require the solution of equations of the second degree or in some cases a greater degree, as the following two examples show:

Make me of 10 such two parts that the one is root of the other.

The problem requires the solution of the system of equations $x + y = 10$; $x = \sqrt{y}$ and is easily traceable to the solution of the second-degree equation $x^2 + x = 10$. In conclusion, we obtain $x = \sqrt{(10 + 1/4)} - 1/2$.

Find me 3 numbers that are in proportion like 2 and 3, and the second like 3 and 4 with respect to the third, and multiply the first by the second, and multiply the result by the third and let the result to be the root of 12.

Muscarello goes on to illustrate the process. It is a matter of calculating the solution of the system $x = 2n$; $y = 3n$; $z = 4n$; $24n^3 = \sqrt{12}$. The value of n (and thus those of x, y, z) can be determined from the solution of the equation $n^6 = 1/48$. It follows that $x^6 = 64/48$ and therefore x is equal to the square root of the cube root of $1 + 1/3$. Similarly, it can be determined that y equals the cube root of $15 + 3/16$ and z equals the cube root of $85 + 1/3$.

To conclude, we recall that in the part devoted to radicals, there is also a problem reminiscent of those presented by Leonardo Pisano in *Liber quadratorum*:

Find me a number that when I add 9 has a root, and when I subtract 9 has a root.

The calculated number is 21 + 1/4. In fact, 30 + 1/4 has root 5 + ½ and 12 + 1/4 has root 3 + ½.[22]

Calculus and the Arts

The final years of the 15th century saw a continuing production of abacus works, but alongside them, there were also interesting episodes of blending traditional themes of computational and geometric mathematics with artistic instances emerging from Italy's humanistic circles. Of these, two notable contributions came from a small town on Tuscany's border with Umbria and Marche: Borgo Sansepolcro. The first concerns a figure whose prominence in the abacus field — compared to his renown in the field of painting — is known to only a few: Piero della Francesca.

Piero, the son of Benedetto de' Franceschi and Romana di Perino di Carlo da Monterchi (1416–1492), known as Piero della Francesca, was born in Borgo Sansepolcro to a family of merchants and, in connection with the family's activities, attended an abacus school as a child.[23] After starting his work as an artist at the workshop of Domenico Veneziano in 1439, he followed his master to Florence, where he came into contact with the humanistic milieu and met not only personalities from the artistic world such as Leon Battista Alberti but also, it seems, Cardinal Bessarione and the mathematician Regiomontanus. After fervid activity that led him to work at major Italian courts in Ferrara, Rome, Arezzo, Urbino, and Rimini, he returned to Borgo Sansepolcro at the age of about sixty, lived there almost continuously (with the exception of a brief period spent in Rimini from 1482 to 1485), and died there on October 12, 1492, the fateful date that by convention separates the medieval world from the modern age.

Piero della Francesca's mathematical output consists in three works: *Trattato d'abaco*, which was written between 1450 and 1460, and two works devoted to geometry and the implications of geometry and optics on painting, *De prospectiva pingendi*, from 1475–1477, and *Libellus de quinque corporibus regularibus*, which appears to have been written

---

[22] This is one of the cases where the text contains an error, because it presents the value 5 + 1/5 instead of the correct value 5 + 1/2.

[23] See Ambrosetti, *op cit.*, pp. 292–299.

during his final stay in Rimini and was dedicated to Guidubaldo di Montefeltro. *Trattato d'abaco* was not written like similar works in an abacus school; it was written for colleagues and friends of the author, themselves merchants or artists, and is described by historian Nadia Ambrosetti as a collective work "containing previous writings brought together without critical revision of the material." It presents 574 problems with their related solutions according to a classical four-part scheme: arithmetic of integers and fractional numbers with related business applications, algebra, geometry, and geometric problems approached using algebraic methods. As always in this type of treatise, the problems are formulated with reference to particular numerical instances and the solution algorithms are given in a prescriptive form — "Do as follows: ..." — without any demonstration of the correctness of the method or proof of the correctness of the solution. This leaves room in various sections of the work for the presentation of incorrect solutions. Nevertheless, the most interesting part of the treatise is the algebraic part, which is particularly detailed because algebra is addressed as both a subject in itself and a method for solving problems of a general type (geometric or arithmetic), so much so that the text might be called a "treatise on algebra".[24] Piero does not limit himself to putting forward the usual problems attributable to first- and second-degree equations; after presenting rules for calculating with radicals and some notable products, he proposes as many as sixty-four cases of equations, some of a general type, including third-, fourth-, fifth-, and sixth-degree equations, with the related solving rules. Unfortunately, as noted earlier, these solving rules (which also seem to have been taken from earlier texts such as Gherardi's, and for which Piero provides no explanation) are not always correct, or rather, they turn out to be correct only in some particular cases.[25]

---

[24] Enrico Gamba, Vico Montebelli and Pierluigi Piccinetti, *La matematica di Piero della Francesca*, in "Lettera Matematica" 59, p. 53.

[25] As we know, general rules for solving third- and fourth-degree equations were not provided until the following century by the Italian mathematicians Scipione del Ferro, Gerolamo Cardano, Niccolò Fontana (known as Tartaglia), and Lodovico Ferrari, while fifth-degree equations had to wait until the 19th century. In the 15th century, various authors were aware of the difficulty of finding a general solving formula for equations of a degree greater than two; see *ibid.*

In the following example we show how the formula provided by Piero for third-degree equations is actually useful for solving a problem concerning the calculation of compound interest over a three-year period:

At what monthly unit rate does 100 *lire* initial give 150 *lire* in 3 years?

Piero proceeds as follows.[26] If x is the monthly interest expressed in *denarii*, the annual interest is x *soldi* per year (remember that 1 *soldo* corresponds to 12 *denari*), so $x/20$ lira per year (remember that 1 *lira* corresponds to 20 *soldi*). At the end of the first year, therefore, the capital has become $M1 = 100(1 + 5x)$; at the end of the second year, it becomes $M2 = 100 + (100 + 5x)x/20 = 100 + 10x + x^2/4$; and at the end of the third year, it will be $M3 = 100 + (100 + 10x + x^2/4)x/20 = 100 + 15x + 3/4\,x^2 + x^3/80$.

Thus, the equation to be solved is $15x + 3/4x^2 + x^3/80 = 50$, that is, $1{,}200x + 60\,x^2 + x^3 = 4{,}000$. For the resolution, Piero uses the formula $x = {}^3\sqrt{[(a/b)^3 + d/c]} - a/b$ with $a = 1{,}200$, $b = 60$, $c = 1$, $d = 4{,}000$ and gets $x = {}^3\sqrt{12{,}000} - 20$.

In addition to solving equations of various degrees, Piero della Francesca's treatise also presents a number of problems of indeterminate analysis, partly derived from Leonardo Pisano's *Liber quadratorum*. The following is one of these problems:

Find me a square number that by subtracting 7 will still be a square and adding 7 will be a square.

It will be recalled that similar problems were mentioned by Leonardo Pisano, who attributed their origin to the mathematical curiosity of Giovanni of Palermo, one of the learned men of Frederick II's court. The problem admits infinite solutions and Piero identifies the following solution:

$$(337/120)^2 - 7 = (113/120)^2$$
$$(337/120)^2 + 7 = (463/120)^2 .$$

Another indeterminate problem is presented in another proposition: Find three square numbers such that the sum of the first two is a square

---

[26] *Ibid.*, p. 55.

and the sum of it with the third is still a square. In modern terms, look for x, y, z such that $x^2 + y^2 = a^2$ and $a^2 + z^2 = b^2$.

Piero starts from the Pythagorean triplet (3, 4, 5) and obtains $x^2 = 9$, $y^2 = 16$ and $a^2 = 25$; then, he places $z^2 = a^2\ 16/9$, obtaining $b^2 = x^2 + y^2 + z^2 = a^2\ (1 + 16/9) = a^2\ (25/9) = (25/3)^2$. Of course, a different solution can be obtained by starting from any of the infinite Pythagorean triplets.

As mentioned earlier, Piero della Francesca's treatise does not present characteristics of originality, but is primarily the result of the transcription of problems from other mathematical texts. More original and more interesting, then, in a broad presentation of computational mathematics in the second half of the 15th century, are *De prospectiva pingendi* and *Libellus de quinque corporibus regularibus*. In these two treatises, which deal with the visual representation of objects and environments, we have an example of the application of computation not to the daily life of merchants but to the artistic activity of painters. It would be foreign to the purposes of this book to discuss the pictorial and aesthetic problems of the late 15th century at this point, but it is nonetheless a well-known fact that the representation of the third dimension was a very prominent topic at the time and one at which Brunelleschi, Alberti, and Paolo Uccello had already tried their hand. In the 1430s, the great humanist, architect, writer, and mathematician Leon Battista Alberti had systematically, albeit still empirically, addressed the issue of perspective in the first volume of *De pictura*. In this work, which is written in both Latin and vernacular, Alberti also illustrated the question of the "circumscription" and "composition" of bodies in a pictorial work, the choice of colors and light, and finally, in the third volume, the figure of the painter himself, who for the first time is presented as an artist and not as a craftsman. Compared with Alberti's work, Piero's *De prospectiva pingendi* takes the problem of perspective important steps forward because he addresses it in what we might call more formal terms by presenting precise definitions, theorems demonstrated in a constructive way, and a series of problems solved by geometric methods. *De prospectiva pingendi* was written in vernacular and was translated into Latin later so that it would take on greater dignity. According to Luca Pacioli, the translator was a certain Maestro Matteo from Borgo Sansepolcro. It is interesting to note how the author precisely defines the purpose of the volume in its first lines by identifying the role and characteristics of measure (*commensuratione*) in a painting: "Painting contains in itself three principal parts, that we say to be design, measure and coloring. By design we mean making profiles and contours that define the subject. By measure we mean having the profiles and contours proportionally displayed in the right place. By coloring we mean giving the colors as they appear in the subject, clear or dark according to the light. Of these three parts, I intend to address only the measure, which we call *prospectiva* (perspective), mixing with it some aspect of design, because without design the perspective cannot be explained; the coloring we will leave aside and we will deal with the part that can be demonstrated with lines, angles, and proportions. And this part contains five items: the first is the vision — that is the eye — the second is the shape of the object seen, the third is the distance between the eye and the object seen, the fourth is the lines that start from the profile of the subject and reach the eye, and the fifth is the position that is between the eye and the subject seen where we intend to place other things."[27]

The work, which shows the author's familiarity with Euclid's texts[28] — particularly *Optics*, cited as *De aspectuum diversitate*, and *Elements* — consists of three books in which the propositions are supported by eighty-two figures. The first book, which is more didactic in nature, deals with plane geometry. It contains exercises related to superficial perspectives in order of increasing difficulty, from the floor grid to some regular polygons (triangle, octagon, hexadecagon, hexagon, and pentagon). As an example, let us see how Piero tackles the representation of two parallel segments, BC and HI, each divided into five sections, setting point A as the vanishing point (Proposition I.8 and related figure):

> Given the line BC, which is divided into D, E, F, G, and another line drawn equidistant to that, which is HI, and from point A let us draw AB, AD, AE, AF, AG, and AC, which divide HI at points K, L, M, N: I say to be divided in that proportion which is the given line BC because BD

[27] Piero della Francesca, *De prospectiva pingendi*, edited by Chiara Gizzi, Edizioni Ca' Foscari, Venice, 2016, p. 81.

[28] According to Vasari, "he was studious in art, and in perspective he was worth so much that no one was more admirable in the things of Euclid's cognition than he"; cf. *ibid.*, p. 15.

is to DE as HK is to KL, and EF to FG is like LM to MN, and FG to GC
is like MN to NI; and the triangle ABD is similar to the triangle AHK,
so ADE is to triangle AKL, and AEF is similar to triangle ALM, so that
they are proportional. And such proportion is between AB and BC and
from AH to HI, and being proportional the major bases so are the minor
bases; and the angles of the triangle ABD are similar to the angles of the
triangle AHK, therefore they are proportional, as is shown in the 21st
proposition of the Sixth Book of Euclid.

In the second book, which deals with the representation of solid
bodies, we also proceed from simpler exercises (a representation of a
cube) to more complex exercises (a temple with an octagonal base, a
cross vault).

In the third book, which contains even more difficult exercises, Piero
first gives his own definition of perspective in the following terms: "I say
that perspective concerns things seen from afar, represented with proportion, according to the amount of their distance, so that things can be correctly degraded (decreased)." The term degraded is then clarified: "But I
say that a degraded proportion is not like 4, 8, 12, 15, or like 6, 9, 11, 12,
but depends on the distance from the eye to the point where the degraded
things are placed."[29] As mentioned earlier, the exercises contained in this
book are even more complex. They involve not only architectural elements such as bases of columns and elaborate capitals but also objects
such as cups and natural elements like heads and human figures.

Unlike what is announced in the title, Piero della Francesca's third
mathematical work, *Libellus de quinque corporibus regularibus*, concerns
not only the properties of the five regular solids (tetrahedron, cube, octahedron, dodecahedron, and icosahedron) but also those of four 'semiregular' or Archimedean solids (truncated cube, truncated octahedron, truncated
dodecahedron, and truncated icosahedron).[30] The text, which was written
in Latin, contains one hundred and forty propositions. The stereometric
projection (the three-dimensional representation) of polyhedra is studied
first, followed by the stereometric projection of architectural structural

---

[29] *Ibid*, pp. 19–20.
[30] Ambrosetti, *op cit.*, p. 297.

elements, columns, capitals, vaults, apses, and so on. Then, more complex problems are tackled, such as calculating the volume of the orthogonal intersection of two cylinders of equal diameter (corresponding to twice the volume of a "pavilion vault").[31]

There is a curious episode concerning the *Libellus* that generally helps us understand how difficult it is to establish the authorship of contributions encountered in medieval mathematical works: It was included by the mathematician Luca Pacioli in his work *Summa de arithmetica, geometria, proportioni et proportionalità* without its source being cited, and so for many years,[32] Pacioli himself was believed to be the author.

This episode leads us to the second character, whose life testifies to the close relationship between the mathematical and artistic worlds in those years. Fra Luca Bartolomeo de Pacioli was a Franciscan friar, a mathematician, and an expert in mercantile matters. He was called Luca di Borgo because he was also a native of Borgo Sansepolcro, where he was born in 1447 and where he died in 1517. After studying Latin, liberal arts, and mathematics at the Rialto school in Venice, he began itinerant work as a teacher of mathematics (including at the university level) in Perugia, Zara, Naples, Rome, Milan, Venice, Padua, and Bologna. In Urbino, he was also tutor to Guidubaldo da Montefeltro. During his peregrinations, Pacioli came into contact with some of the leading figures in the artistic world of the time — Leonardo da Vinci, Piero della Francesca, Melozzo da Forlì, Bramante, and perhaps Albrecht Dürer himself. Despite his eventful life, Luca Pacioli had the time to write several mathematical works of considerable interest. First of all, between 1477 and 1478, he

---

[31] The results presented by Piero seem to indicate that he was familiar with Greek mathematical works such as Archimedes' *The Method* or Heron's *Metrics,* but there is no certainty of this. What is certain is that the subject of vaults was already being studied among architects (as can be verified in Leon Battista Alberti's *De re aedificatoria*), although from the constructive rather than mathematical point of view; see Gamba, Montebelli and Piccinetti, *op cit.*, p. 51.

[32] The fact that part of *Divina Proportione* was based on an earlier work by Piero della Francesca was only discovered in 1915 by Girolamo Mancini; see Enrico Giusti, *Il Rinascimento. Verso una nuova matematica*, 2001 (https://www.treccani.it/ encyclopedia/ il-rinascimento-verso-una-nuova-matematica).

wrote *Tractatus mathematicus ad discipulos Perusinos*[33] for his students in Perugia. Subsequently, in 1494, he published his first important work, *Summa de arithmetica, geometria, proportioni et proportionalità*, a text that took up and extended what he had written for his students in Perugia and which he dedicated to Guidubaldo.

After a series of journeys (some of which he made in the company of Leonardo, whom he had met in Milan when he was a guest of Ludovico il Moro), he returned to Venice at the beginning of the 16th century, where in 1509 he had the three volumes of *De divina proportione* printed, to whose iconography Leonardo himself contributed. Other works by Luca Pacioli that remained unfinished after his death were devoted to recreational mathematics (*De viribus quantitatis*) and the game of chess (*De ludo scachorum* or *Schifanoia*,[34] a work that was thought to have been lost and was only found in 2006). *Summa*[35] is considered by historians to be a work that has been "more influential than original"[36] because it is a collection of contributions by earlier authors — Euclid, Boethius, Leonardo Pisano, and Giordano Nemorario — whom, moreover, Luca Pacioli does not generally quote. The work is divided into two parts: The first, consisting of nine "distinctions", is devoted to arithmetic, algebra, and commercial mathematics, while the second is devoted to geometry. The distinctions dealing with arithmetic present fundamental algorithms for operations with integer and fractional numbers, the rule of three, and the theory of proportions. The seventh and eighth distinctions, which are devoted to algebra and inspired by *Liber abaci*, present the methods of false position (simple and double) and the canonical rules for solving first- and second-degree equations. They also show how numerous other cases of higher degrees can be generated, although for many of them, the author is unable to provide resolution rules (Pacioli believed that even third-degree

---

[33] Francesca Aceto, *Scienza e gioco nel* De viribus quantitatis *di Luca Pacioli*, in *Incontri di Studio MAES*, 2008, p. 49.

[34] In the dedicatory epistle of *De viribus quantitatis*, the text is called *De Ludis* and is presented as "an entertaining and joyful treatise *de ludis* in general, especially about the game of chess, called Schifanoia, with disapproval of illicit games."

[35] Ambrosetti, *op cit.*, pp. 300–303.

[36] Boyer, *Storia della matematica*, p. 323.

equations were not in general solvable by using algebraic methods).[37] Finally, the ninth distinction (which contains *Tractatus particularis de computis et scripturis*) has become famous for its presentation of double-entry bookkeeping. A final observation concerns the style of presentation of problems and algorithms: Luca Pacioli uses syncopated notation with various abbreviations. For example, to indicate *cosa* (thing, the unknown appearing in the equation), he writes *co*; for *censo* (value, the square of the unknown) *ce*; for equality *ae*; and to represent the root, he uses the symbol *R*.

The geometric section of *Summa* is not particularly significant. More geometrically stimulating, in part thanks to the illustrations by Leonardo, is Luca's other famous work, *De divina proportione*. In it, regular and semiregular polygons and polyhedra are studied, and more generally mathematical proportion is illustrated, particularly the so-called golden ratio and its applications in architecture. As already mentioned, the Italian text of *Libellus de quinque corporibus regularibus* is also included in the work with no citation by the author.

There are also some very interesting aspects of computational mathematics in *De viribus quantitatis* (a title we might translate as "On the power of numbers"), which has one of the largest collections of mathematical games and recreational problems and was written between 1496 and 1509. This manuscript was also part of Prince Baldassarre Boncompagni's collection, and it is now preserved in the University Library in Bologna. The text consists of three parts: One is devoted to "strength of numbers", that is, games based on arithmetic properties; and one is devoted to "geometric strength", in which exercises and problems solvable with a ruler and compass are proposed. The third part, conversely, is extremely heterogeneous[38] and contains games of a logical type, including classic puzzles that we have already seen (such as the one about 3 wives and 3 jealous husbands, which appears in Chapter LXI); proverbs of popular wisdom; topics and advices of mercantile interest;

---

[37] *Ibid.*, p. 324.

[38] An extensive series of examples of the games and puzzles contained in Luca Pacioli's text is presented in Vanni Bossi, Antonietta Mira, and Francesco Arlati, *Mate-magica. I giochi di prestigio di Luca Pacioli*, Aboca, Sansepolcro, AR, 2012.

recipes; jokes; pastimes; juggling with ropes, coins, glasses, needles, nails, and eggs (see Chapter LII: "How to make an egg stand with no help"[39] or Chapter CXXIIII: "Cut an apple inside without cutting the peel and similarly a peach, an orange etc."); and sensory, optical, or tactile illusions. Finally, Pacioli presents problems and *aenigmata* for learned people (more sophisticated and written in Latin) and problems addressed to idiots: that is, games whose success is based on the naiveté of the people to whom they are proposed.[40]

Many of the games presented in *De viribus quantitatis* are taken from classical texts (Alcuin, Leonardo Pisano, Paolo dell'Abbaco, the Calandri brothers, etc.), such as those involving decanting liquids into containers of set capacity until the desired amount of liquid is obtained, which, as we will recall, were proposed by Paolo dell'Abbaco. Among the games that are presented in several versions and that denote Luca Pacioli's undoubted skill with using algorithmic techniques are those consisting in guessing a number thought up by a bystander.[41] A simple example is as follows:

Thirteenth effect: to find a number in all ways.

The number thought up, z, integer or fractional, is broken down by the bystander into the sum of as many parts you wish, integer or fractional, equal to each other or not. Each part is to be multiplied by the starting number z, and then the results are to be added together. If the parts are $x1$, $x2$, ... , xn, the result is $z(x1 + x2 + \cdots + xn) = z^2$. So, the one who has to guess the number z only has to calculate the square root of the result.

The next example is also based on elementary algebraic properties:

Eleventh effect: to find a number in all ways.

In this case, the formula for the square of a binomial is used (but the technique can evidently be generalized to any power, although in Luca

---

[39] The game corresponds to the well-known 'egg of Columbus'.

[40] Aceto, *op cit.*, p. 52.

[41] *Ibid.*, pp. 33–35.

Pacioli's time, the general formula based on Tartaglia's triangle may not have been known). Given a number z, one is therefore asked to do the following:

- divide it into two addends x and y such that $x + y = z$;
- calculate x^2, y^2, $2xy$ and add them up, getting the result $t = (x + y)^2 = z^2$.

Again, to guess the number z thought of by the bystander, it is enough to calculate the square root of t. Here, Luca Pacioli makes explicit reference to Euclid ("And this rule can be derived from the 4th rule of the second book of our philosopher").

A third example is based on a formula that was also used by Fibonacci: $[z (z + 1) - (2x + zy)] = t = x (z - 1) + y$.

First effect: divide a number in two parts.[42]

Given a known integer z, two bystanders are asked to think of two values x and y such that $x + y = z$. The game consists of guessing the two parts x and y. One of the two bystanders is asked to multiply his part by 2 and the other to multiply his part by z. They are then asked to add the two values obtained and subtract them from the $z(z + 1)$ value, obtaining the value t that is revealed. At this point, the one who leads the game divides t by $z - 1$; the quotient of the division will be the x value and the remainder will be the y value. It should be noted, however, that the trick works correctly if dealing with integer values and if x and y are different from 1.

In conclusion, it should be emphasized that Luca Pacioli's works highlight a witty personality with vast cultural knowledge, capable of ranging across different fields of mathematics and stimulating his readers' curiosity and interest. *Summa* in particular is unanimously viewed as a valuable work because, as the title says, it represents an encyclopedic compendium of knowledge in the fields of arithmetic, algebra, geometry, trigonometry, and commercial mathematics, and is thus one of the cornerstones of medieval mathematics after *Liber abaci*.

---

[42] *Ibid.*, p. 22.

Conclusions

With the presentation of the works of Luca Pacioli, a mathematician who played an important role at the turn of the 15th and 16th centuries, we conclude these pages, in which we aimed to trace the presence of algorithms in daily life in the Middle Ages. We have shown how, especially between the 9th and 15th centuries, the close relationship between algorithms and the problems of everyday life helped — first in the Islamic world and then in Europe — to promote the development of computational mathematics. As we have seen, after computational methods based on the Indo-Arab positional numbering system became established in the West and the works of leading Arabic mathematicians were popularized, arithmetic and algebraic computations were given a major boost, mainly thanks to the abacus schools where merchants' children learned the basic algorithms for buying and selling goods, exchanging currencies, calculating investment costs and profits, and dividing inheritances.

From our modern-day perception, the term 'algorithm' sends us back to a presentation of rigorous and formalized computational rules. Apparently the formulation of the computational methods proposed in the texts we have dealt with seems to be quite a long way from a style of presentation such as this. However, while their aim was to solve particular numerical cases, these computational methods almost always had a very detailed didactic explanatory character and a precision that made it possible to reapply the same methods to different numerical cases; when taken all together, they have a true algorithmic connotation. Today, abacus books and books of "algorisms", with their wealth of fundamental mathematic concepts and applied examples, constitute an enormous heritage for both the study of the history of mathematics and knowledge of the medieval world, the goods that were handled, the coins that were used, the cities where merchants moved during the fairs, and so on.

Our look at the role of algorithms in the daily life of the Middle Ages ends at the end of the 15th century, the period when a change is noted in the way the study of mathematics is approached. Among the various causes of this evolution that can be identified, two seem to us to be decidedly prevalent. The first is the revival of Greek mathematics. Alongside the continuation of the study of algebra, which led to the definition of methods for the general solution of third- and fourth-degree equations in the following century, the humanistic tension toward the rediscovery of the Greek and Latin classics that fueled Italian (and more generally European) thinkers and intellectuals of the 15th century leads to a reevaluation of the work of mathematicians of antiquity, such as Archimedes, Euclid, Apollonius, and Heron, and those of late antiquity, such as Diophantus, Nicomachus of Gerasa, and Boethius. The effect of this rediscovery obscures the fundamental role that mathematical contributions of Indo-Arabic origin had played in previous centuries. The contributions of not only al-Khwarizmi but also Fibonacci fell into oblivion for a number of centuries, while the abstract and almost philosophical approach to the study of mathematics was reevaluated, according to an approach that is well represented by a passage from Plato's *Repubblica,* in which he argues that the science of numbers "is to be taught not in the vulgar manner, dealing with it for the purpose of buying and selling, like merchants and dealers, but in such a way that the intelligence may contemplate the nature of numbers."[43]

The second event that contributed to changing the context in which mathematical thought was evolving, perhaps even more decisively, was the invention of printing in the mid-15th century. According to the mathematical historian Enrico Giusti, "At the beginning of the sixteenth century, the descending line of Arab science and the ascending line of European mathematics were on the verge of intersecting, when an unpredictable phenomenon came to completely revolutionize the scientific environment. The press [...] took over scientific culture, causing revolutionary effects."[44]

Indeed, printing offered a possibility for the dissemination of works that led to the separation of the production of mathematical treatises and their use by scholars or mere connoisseurs from the sphere in which they had previously been produced: the abacus schools. This development offered mathematical thought a much wider space for dissemination that was less closely linked to applications of interest by mercantile communities. A testimony to the gap the diffusion of printed texts created with

---

[43] Nasti, *L'algorismo nell'aritmetica del Trecento e del Quattrocento*, p. 277.
[44] Giusti, *Il Rinascimento. op cit.*

respect to the previous modes of circulation of manuscripts is provided by the relationship between the work of Piero della Francesca and that of Luca Pacioli; the latter was well aware of the advantages offered by book printing and had no qualms about incorporating texts by other authors, including Piero himself, into his writings.

The effect of these changes was not immediately explicit, but it was instrumental in the emergence of new currents of mathematical thought in the 16th century that moved away from computational mathematics and toward more abstract and more powerful forms such as those developed by Cardano, Tartaglia, and especially Viète, the mathematician who is considered to be the true initiator of modern algebra.[45]

---

[45] See Jeffrey A. Oaks, *François Viète's Revolution in Algebra*, in Archive for History of Exact Sciences, vol. 72, no. 3, Springer, Berlin, 1918.

References

Aceto, Francesca, *Scienza e gioco nel* De viribus quantitatis *di Luca Pacioli*, in *Incontri di Studio del MAES*, 2008. Aczel, Amir D., *Fermat's Last Theorem*, Delta, New York, NY, 1996.

Aïssani, Djamil, Les mathématiques à Bougie médieval et Fibonacci, in *Marcello Morelli e Marco Tangheroni (a cura di)*, *Leonardo Fibonacci*, Pacini Editore, Ospedaletto (PI), 1994.

Al-Khalili, Jim, *La casa della saggezza*, Bollati Boringhieri, Torino, 2010.

Alexander de Villa Dei, Carmen de Algorismo, in James Orchard Halliwell (a cura di), *Rara Mathematica, or a Collection of Treatises on the Mathematics*, Londra, 1841.

Allard, André, Les sources arithmétiques et le calcul indien dans le Liber abaci, in Marcello Morelli e Marco Tangheroni (a cura di), *Leonardo Fibonacci*, Pacini Editore, Ospedaletto (PI), 1994.

Ambrosetti, Nadia, Algorithmic in the 12th Century: The Carmen de Algorismo by Alexander de Villa Dei, in *3rd International Conference on History and Philosophy of Computing (HaPoC)*, Ottobre, 2015, Pisa (10.1007/978-3-319-47286-7_5.hal-01615308).

Ambrosetti, Nadia, *L'eredità arabo-islamica nelle scienze e nelle arti del calcolo dell'Europa medievale*, LED, Milano, 2008.

Anonimo, *Intorno al* Tractatus de abaco *di Gerlando*, in *Bullettino di bibliografia e di storia delle scienze matematiche e fisiche*, tomo X, Roma, 1877.

Anonimo, *Larte de labbacho*, 1478 (www.centromorin.it/info/abacho/abba cho.pdf).

Antoni, Tito, Leonardo Pisano detto il Fibonacci e lo sviluppo della contabilità mercantile del duecento, in Marcello Morelli e Marco Tangheroni (a cura di), *Leonardo Fibonacci*, Pacini Editore, Ospedaletto (PI), 1994.

Arrighi, Gino, Considerazioni sul Liber abaci di Leonardo Pisano, in Marcello Morelli e Marco Tangheroni (a cura di), *Leonardo Fibonacci*, Pacini Editore, Ospedaletto (PI), 1994.

Ausiello, Giorgio e Petreschi, Rossella (a cura di), *The Power of Algorithms*, Springer, Berlino, 2013.

Bagni, Giorgio T., Larte de labbacho (l'Aritmetica di Treviso, 1478) e la matematica medievale, in *I Seminari dell'Umanesimo Latino 2001-2002*, Fondazione Cassamarca, Treviso, 2002 (www.syllogismos.it/history/uma -abbaco.pdf).

Balducci Pegolotti, Francesco, *La pratica della mercatura*, a cura di Allan Evans, in *The Mediaeval Academy of America No. 24, Mediaeval Academy Books*, Cambridge, 1936.

Boretius, Alfred (a cura di), *Epistola de litteris colendis*, in *MGH Capitularia regum Francorum*, vol. I, Hahnsche Buchhandlung, Hannover, 1883.

Bossi, Vanni, Mira, Antonietta e Arlati, Francesco, *Mate-magica. I giochi di prestigio di Luca Pacioli*, Aboca, Sansepolcro (AR), 2012.

Boyer, Carl B., *Storia della matematica*, Mondadori, Milano, 1982.

Brown, Nancy Marie, *The Abacus and the Cross: The Story of the Pope Who Brought the Light of Science to the Dark Ages*, Basic Books, New York, NY, 2010.

Bubnov, Nicolaus (a cura di), *Gerberti Postea Silvestri II Papae Opera Mathematica (972-1003)*, Friedlander, Berlino, 1899.

Burkholder, Peter J., Alcuin of York's "Propositiones ad Acuendos Iuvenes" ("Propositions for Sharpening Youths"), Introduction and Commentary, in *Electronic Bulletin for the History and Philosophy of Science and Technology*, 1, n. 2, 1993.

Calandri, Pier Maria, *Tractato d'abbacho*, a cura e con introduzione di Gino Arrighi, Domus Galilaeana, Pisa, 1974.

Cartocci, Alice, *La matematica degli Egizi. I papiri matematici del Medio Regno*, Firenze University Press, Firenze, 2007.

Chabert, Jean-Luc, *Histoire d'algorithmes. Du caillou à la puce*, Belin, Parigi, 1994.

B. Codenotti, G. Resta, *Rithmomachia. Alla riscoperta di un gioco medievale*, Messaggerie Scacchistiche, 2022

Cotrugli, Benedetto, *Libro dell'arte della mercatura*, a cura di Vera Ribaudo, Edizioni Ca' Foscari, Venezia, 2016 (https://phaidra.cab.unipd.it/detail/o:432210?mycoll=o:432598).

de Fermat, Pierre, *Osservazioni su Diofanto*, Bollati Boringhieri, Torino, 2006.

Devlin, Keith, *I numeri magici di Fibonacci*, Rizzoli, Milano, 2012.

Diem, Albrecht, The Emergence of Monastic Schools. The Role of Alcuin, in *Proceedings of the Third Germania Latina Conference*, Gröningen, 1995.

Djebbar, Ahmed, *Storia della scienza araba. Il patrimonio intellettuale dell'Islam*, Raffaello Cortina Editore, Milano, 2002.

Erodoto, *Storie*, BUR, Milano, 2009.

Folkerts, Menso, *"Boethius" Geometrie II. Ein Mathematisches Lehrbuch des Mittelalters*, Franz Steiner, Stoccarda, 1970.

Franci, Raffaella (a cura di), *Alcuino di York. Giochi matematici alla Corte di Carlomagno*, ETS, Pisa, 2016.

Franci, Raffaella, *Il* Liber abaci *di Leonardo Fibonacci 1202-2002*, in *Bollettino dell'Unione Matematica Italiana*, serie VIII, vol. V-A, 2002.

Franci, Raffaella e Toti Rigatelli, Laura, Maestro Benedetto da Firenze e la storia dell'algebra, in *Historia Mathematica*, 10, 1983.

Gamba, Enrico, Montebelli, Vico e Piccinetti, Pierluigi, La matematica di Piero della Francesca, in *Lettera matematica*, 59 (https://urbinoelapro spettiva. uniurb.it/wp-content/uploads/2017/02/LM59_09.pdf).

Garrison, Mary, The library of Alcuin's York, in Richard Gameson (a cura di), *Cambridge History of the Book in Britain*, vol. I, Cambridge University Press, Cambridge, 2019.

Gavagna, Veronica, Leonardo Fibonacci, in Antonio Clericuzio e Saverio Ricci (a cura di), *Enciclopedia italiana di scienze, lettere ed arti. Il contributo italiano alla storia del pensiero — Scienze*, Istituto della Enciclopedia Italiana, Roma, 2013.

Gengaro, Maria Luisa, Le miniature del codice, in Pietro Paolo Muscarello, *Algorismus. Trattato di aritmetica pratica e mercantile del secolo XV*, a cura di Giorgio Chiarini, Banca Commerciale Italiana, Milano, 1972.

Genocchi, Angelo, Bibliografia del Principe Boncompagni, in *Annali di Scienze Matematiche e Fisiche*, tomo VIII, Roma, 1857.

Geronimi, Nando (a cura di), *Giochi matematici del medioevo*, Bruno Mondadori, Torino, 2006.

Gherardi, Paolo, *Opera matematica. Libro di ragioni — Liber habaci*, a cura di Gino Arrighi, MPF, Lucca, 1987.

Giusti, Enrico, *Il Rinascimento. Verso una nuova matematica*, 2001 (https://www. treccani.it/enciclopedia/il-rinascimento-verso-una-nuova-matematica).

Giusti, Enrico, Matematica e commercio nel Liber abaci, in Enrico Giusti e Raffaella Petti (a cura di), *Un ponte sul Mediterraneo. Leonardo Pisano, la scienza araba e la rinascita della matematica in Occidente*, Polistampa, Firenze, 2002.

Høyrup, Jens, Jordanus de Nemore, 13th Century Mathematical Innovator: An Essay on Intellectual Context, Achievement, and Failure, in *Archive for History of Exact Sciences*, 38, 1988.

Høyrup, Jens, Leonardo Fibonacci and Abbaco Culture. A Proposal to Invert the Roles, in *Revue d'histoire des mathématiques*, 11, 2005.

Høyrup, Jens, The Algorithm Concept — Tool for Historiographic Interpretation or Red Herring?, in *CiE 2008, Logic and Theory of Algorithms*, LNCS 5028, Springer, Berlino, 2008.

Ifrah, Georges, *Histoire universelle des chiffres*, 2 voll., Laffont, Parigi, 1994 (*Storia universale dei numeri*, Mondadori, Milano, 1989).

Jacopo da Firenze, *Tractatus algorismi*, a cura di Jens Høyrup, Roskilde University, Roskilde, 2007.

Jagadguru Swami Sri Bharati Krsna Tirthaji Maharaja, *Vedic Mathematics*, Motilal Banarsidass Pub., Delhi-Varanasi-Patna, 1989.

Joannis de Sacro-Bosco, Tractatus de arte numerandi, in James Orchard Halliwell (a cura di), *Rara Mathematica, or a Collection of Treatises on the Mathematics*, Londra, 1841.

Jordanus de Nemore, *De numeris datis. A critical edition and translation by Barnabas Hughes*, University of California Press, Berkeley, CA, 1981.

Knuth, Donald E., Ancient Babylonian Algorithms, in *Communications of the ACM*, 15, 7, 1972.

Koetsier, Teun e Bergmans, Luc (a cura di), *Mathematics and the Divine. A Historical Study*, Elsevier, Amsterdam-Filadelfia, 2005.

Kramer, Rutger, 'Ecce Fabula!' Problem-Solving by Numbers in the Carolingian World: The Case of the Propositiones ad Acuendos Iuvenes, in *Proceedings of the 2015 MEMSA Student Conference*, 2015.

Laura, Luigi, *Breve ed universale storia degli algoritmi*, Luiss University Press, Roma, 2019.

Leonardo de Pisa, *El libro de los números cuadrados*, introduzione di Paul Ver Eecke, EUDEBA, Buenos Aires, 1973.

Leonardo Pisano, *Il libro dei quadrati*, introduzione e commenti di Ettore Picutti, in *Physis*, vol. XXI, 1979.

Maccagni, Carlo (a cura di), *Atti del primo convegno internazionale di ricognizione delle fonti per la storia della scienza italiana: i secoli XIV-XVI*, Barbèra Editore, Firenze, 1967.

Migne, Jacques-Paul (a cura di), *B. Flacci Albini seu Alcuini abbatis et Caroli magni imperatoris magistri opera omnia*, 2 voll., Parigi, 1851 (ristampa Hachette BnF, Parigi, 2018).

Moyer, Ann E., *The Philosopher's Game: Rithmomachia in Medieval and Renaissance Europe, with an Edition of Ralph Lever and William Fulke — The Most Noble, Ancient and Learned Playe (1563)*, The University of Michigan Press, Ann Arbor, MI, 2001.

Muscarello, Pietro Paolo, *Algorismus. Trattato di aritmetica pratica e mercantile del secolo XV*, a cura di Giorgio Chiarini, Banca Commerciale Italiana, Milano, 1972.

Nasti, Mauro, L'algorismo nell'aritmetica del Trecento e del Quattrocento, in Pietro Paolo Muscarello, *Algorismus. Trattato di aritmetica pratica e mercantile del secolo XV*, a cura di Giorgio Chiarini, Banca Commerciale Italiana, Milano, 1972.

Nemet-Nejat, Karen Rhea, *Cuneiform Mathematical Texts as a Reflection of Everyday Life in Mesopotamia*, American Oriental Society, New Haven, CT, 1993.

Oaks, Jeffrey A., François Viète's Revolution in Algebra, in *Archive for History of Exact Sciences*, vol. 72, n. 3, Springer, Berlino, 1918.

Paolo dell'Abbaco, *Le regoluzze di Maestro Paolo dell'Abbaco, matematico del secolo XIV*, Fondazione Datini, 1860.

Paolo dell'Abbaco, *Trattato d'aritmetica. Secondo la lezione del Codice Magliabechiano XI, 86 della Biblioteca Nazionale di Firenze*, a cura di Gino Arrighi, Domus Galilaeana, Pisa, 1964.

Pegolotti, Francesco Balducci, *La pratica della mercatura*, a cura di Allan Evans, in *The Mediaeval Academy of America*, 24, Cambridge, 1936.

Pergola, Ruggiero, *Ex arabico in latinum: Traduzioni scientifiche e traduttori nell'Occidente medievale*, in "Studi di Glottodidattica", 3, 2009.

Piero della Francesca, *De prospectiva pingendi*, a cura di Chiara Gizzi, Edizioni Ca' Foscari, Venezia, 2016.

Poole, Reginald Lane, *The Exchequer in the Twelfth Century*, Oxford, 1912.

Posamentier, Alfred S. e Lehmann, Ingmar, *I (favolosi) numeri di Fibonacci*, Muzzio, Roma, 2010.

Pratesi, Franco, Il gioco dei filosofi fiorentini, in *L'Italia Scacchistica*, n. 7, 1995.

Puri, Narinder, *An Overview of Vedic Mathematics*, Rajasthan University, Jaipur, 1988.

Rashed, Roshdi, *Al-Khwarizmi. The Beginning of Algebra*, Saqi, Londra, 2009.

Rashed, Roshdi, L'algèbre, in Id. (a cura di), *Histoire des sciences arabes, Mathématique et physique*, vol. II, Seuil, Parigi, 1997.

Riché, Pierre, *Gerbert d'Aurillac. Le pape de l'an mil*, Fayard, Parigi 2006. Richero di Saint Remi, *I quattro libri delle Storie (888-998)*, Pisa University Press, Pisa, 2008.

Rossi, Paolo, Algoritmi matematici nelle lettere di Gerbert, in *Gerbertus*, vol. 1, 2010.

Rucquoi, Adeline, *Histoire médiévale de la Péninsule ibérique*, Seuil, Parigi, 1993.

Russo, Lucic, *La rivoluzione dimenticata*, Feltrinelli, Milano, 1998.

Sapori, Armando, La cultura del mercante medievale italiano, in Gabriella Airaldi (a cura di), *Gli orizzonti aperti. Profili del mercante medievale*, Scriptorium, Torino, 1997.

Sapori, Armando, *La mercatura medievale*, Sansoni, Firenze, 1972.

Sapori, Armando, *Studi di storia economica. Secoli XIII-XIV-XV*, vol. I, Sansoni, Firenze, 1982.

Sepkoski, David, Ann E. Moyer: The Philosopher's Game: Rithmomachia in Medieval and Renaissance Europe, in *Isis*, vol. 95, n. 4, 2004.

Sesiano, Jacques, Islamic Mathematics, in Helaine Selin (a cura di), *Mathematics Across Cultures: The History of Non-Western Mathematics*, Springer, Berlino, 2001.

Sigismondi, Costantino, La sfera di Gerberto, in *Gerbertus*, vol. 1, 2010.

Sigler, Laurence E., *Fibonacci's Liber abaci*, Springer, Berlino, 2003.

Silva, Jorge Nuno, Teaching and Playing 1000 Years Ago. Rithmomachia, in Costantino Sigismondi (a cura di), *Orbe Novus. Astronomia e studi gerbertiani*, Universitalia, Roma, 2010.

Singmaster, David e Hadley, John, Problems to Sharpen the Young, in *The Mathematical Gazette*, vol. 76, n. 475, 1992.

Smith, David Eugene e Eaton, Clara C., Rithmomachia, the Great Medieval Number Game, in *The American Mathematical Monthly*, vol. XVIII, n. 4, 1911.

Spence, Jonathan D., *Il palazzo della memoria di Matteo Ricci*, Adelphi, Milano, 2010.

Starr, Frederick S., *Lost Enlightenment: Central Asia's Golden Age from Arab Conquest to Tamerlane*, Princeton University Press, 2013.

Suzuki, Jeff, *Mathematics in Historical Context*, Mathematical Association of America, 2009.

Swetz, Frank J., *Capitalism and Arithmetic: The New Math of the 15th Century*, Including the Full Text of the Treviso Arithmetic of 1478, Open Court, La Salle, IL, 1987.

Sylla, Edith D., Book Review: De numeris datis. Jordanus de Nemore, Barnabas Bernard Hughes, in *Isis*, vol. 74, n. 1, 1983.

Tangheroni, Marco, Fibonacci, Pisa e il Mediterraneo, in Marcello Morelli e Marco Tangheroni (a cura di), *Leonardo Fibonacci*, Pacini Editore, Ospedaletto (PI), 1994.

Thiery, Antonio, Federico II e le scienze, in Angiola Maria Romanini (a cura di), *Problemi di metodo per la lettura dell'arte federiciana. Federico II e l'arte del Duecento italiano*, vol. II, Congedo Editore, Galatina (LE), 1980.

Tigar, Michael E., *Law and the Rise of Capitalism*, Monthly Review Press, New York, NY, 2000.

Travaini, Lucia, *Monete, mercanti e matematica*, Jouvence, Sesto San Giovanni (MI), 2003.

Treutelein, Pietro, Intorno ad alcuni scritti inediti relativi al calcolo dell'abaco, in *Bullettino di bibliografia e di storia delle scienze matematiche e fisiche*, tomo X, Roma, 1877.

Tucci, Ugo, Manuali d'aritmetica e mentalità mercantile tra Medioevo e Rinascimento, in Marcello Morelli e Marco Tangheroni (a cura di), *Leonardo Fibonacci*, Pacini Editore, Ospedaletto (PI), 1994.

Ulivi, Elisabetta, *Gli abacisti fiorentini delle famiglie 'del Maestro Luca', Calandri e Micceri e le loro scuole d'abaco*, Olschki, Firenze, 2013.

Ulivi, Elisabetta, Scuole e maestri d'abaco in Italia tra Medioevo e Rinascimento, in Enrico Giusti e Raffaella Petti (a cura di), *Un ponte sul Mediterraneo. Leonardo Pisano, la scienza araba e la rinascita della matematica in Occidente*, Polistampa, Firenze, 2002.

Unali, Anna, *Le fil rouge entre l'Europe et l'Afrique, Accords commerciaux et politiques de conquête (XIe-XVe siècle)*, L'Harmattan, 2023.

Veratti, Bartolomeo, *De' matematici italiani anteriori all'invenzione della stampa. Commentario storico*, Modena, 1840.

Warntjes, Immo, Introduction: State of Research on Late Antique and Early Medieval Computus, in Immo Warntjes e Dáibhí Ó. Cróinín (a cura di), *Late Antique Calendrical Thought and its Reception in the Early Middle Ages*, Brepols, Turnhout, 2017.

Youschkevitch, Adolf P., *Les mathématiques arabes*, Vrin, Parigi, 1976.

Zuccato, Marco, Gerbert of Aurillac and a Tenth-Century Jewish Channel for the Transmission of Arabic Science to the West, in *Speculum*, vol. 80, n. 3, 2005.